Forestry: Afforestation and Deforestation

Forestry: Afforestation and Deforestation

Edited by
Roland Graves

Larsen & Keller
www.larsen-keller.com

Forestry: Afforestation and Deforestation
Edited by Roland Graves
ISBN: 978-1-63549-692-5 (Hardback)

☰ Larsen & Keller

Published by Larsen and Keller Education,
5 Penn Plaza,
19th Floor,
New York, NY 10001, USA

Cataloging-in-Publication Data

Forestry : afforestation and deforestation / edited by Roland Graves.
    p. cm.
Includes bibliographical references and index.
ISBN 978-1-63549-692-5
1. Forests and forestry. 2. Afforestation. 3. Deforestation. I. Graves, Roland.
SD373 .F67 2018
634.9--dc23

For more information regarding Larsen and Keller Education and its products, please visit the publisher's website www.larsen-keller.com

Table of Contents

Preface

Many actions of human beings have led to deforestation which has resulted in loss of biodiversity and natural habitats. Therefore, forestry as a subject is very important, as it deals with process of creating, managing, repairing, covering forests and protecting other environmental habitats. Afforestation or reforestation refers to the process of planting trees in a space which was left barren before. This book mainly focuses on the subject of forestry, with special significance on the areas of deforestation and afforestation. The topics introduced in it cover many new and important theories as well as practices under this field. This textbook is an essential guide for both academicians and those who wish to pursue this discipline further.

Given below is the chapter wise description of the book:

Chapter 1- The science of conserving and managing forests is known as forestry. Modern forestry focuses on concerns such as conservation of wildlife habitats, provision of timber, water quality management and community management. This is an introductory chapter which will introduce briefly all the significant aspects of forestry.

Chapter 2- Afforestation is the growing of trees in areas where there were previously no trees. Government organizations and non-profitable organizations engage in the practice of afforestation. This chapter is an overview of the subject matter incorporating all the major aspects of afforestation.

Chapter 3- Deforestation is the clearing of forests to convert them into farms, ranches and other non-forest areas. Apart from human interference, the natural causes of deforestation are drought, soil erosion and flash flood. Desertification, habitat destruction, global warming, arctic sea ice decline and future sea level are some of the topics discussed in this section. The aspects elucidated in this chapter are of vital importance, and provide a better understanding of forestry.

Chapter 4- Forest protection is the conservation of forest regions that are under threat. Unsustainable farming, pollution of soil and urban sprawl are some of the causes of forest depletion. Forest protection and management can best be understood in confluence with the major topics listed in the following chapter. Forestry is best understood in confluence with the major topics listed in the following chapter.

Indeed, my job was extremely crucial and challenging as I had to ensure that every chapter is informative and structured in a student-friendly manner. I am thankful for the support provided by my family and colleagues during the completion of this book.

Editor

Defining Forestry

The science of conserving and managing forests is known as forestry. Modern forestry focuses on concerns such as conservation of wildlife habitats, provision of timber, water quality management and community management. This is an introductory chapter which will introduce briefly all the significant aspects of forestry.

Forest

A conifer forest in the Swiss Alps (National Park)

A forest is a large area dominated by trees. Hundreds of more precise definitions of forest are used throughout the world, incorporating factors such as tree density, tree height, land use, legal standing and ecological function. According to the widely used Food and Agriculture Organization definition, forests covered four billion hectares (15 million square miles) or approximately 30 percent of the world's land area in 2006.

Forests are the dominant terrestrial ecosystem of Earth, and are distributed across the globe. Forests account for 75% of the gross primary productivity of the Earth's biosphere, and contain 80% of the Earth's plant biomass.

Forests at different latitudes and elevations form distinctly different ecozones: boreal forests near the poles, tropical forests near the equator and temperate forests at mid-latitudes. Higher elevation areas tend to support forests similar to those at higher latitudes, and amount of precipitation also affects forest composition.

Human society and forests influence each other in both positive and negative ways. Forests provide ecosystem services to humans and serve as tourist attractions. Forests can also affect people's health. Human activities, including harvesting forest resources, can negatively affect forest ecosystems.

Definition

Forest in the Scottish Highlands

Although *forest* is a term of common parlance, there is no universally recognised precise definition, with more than 800 definitions of forest used around the world. Although a forest is usually defined by the presence of trees, under many definitions an area completely lacking trees may still be considered a forest if it grew trees in the past, will grow trees in the future, or was legally designated as a forest regardless of vegetation type.

There are three broad categories of forest definitions in use: administrative, land use, and land cover. Administrative definitions are based primarily upon the legal designations of land, and commonly bear little relationship to the vegetation growing on the land: land that is legally designated as a forest is defined as a forest even if no trees are growing on it. Land use definitions are based upon the primary purpose that the land serves. For example, a forest may be defined as any land that is used primarily for production of timber. Under such a land use definition, cleared roads or infrastructure within an area used for forestry, or areas within the region that have been cleared by harvesting, disease or fire are still considered forests even if they contain no trees. Land cover definitions define forests based upon the type and density of vegetation growing on the land. Such definitions typically define a forest as an area growing trees above some threshold. These thresholds are typically the number of trees per area (density), the area of ground under the tree canopy (canopy cover) or the section of land that is occupied by the cross-section of tree trunks (basal area). Under such land cover definitions, and area of land only be defined as forest if it is growing trees. Areas that fail to meet the land cover definition may be still included under while immature trees are establishing if they are expected to meet the definition at maturity.

Under land use definitions, there is considerable variation on where the cutoff points are between a forest, woodland, and savanna. Under some definitions, forests require very high levels of tree canopy cover, from 60% to 100%, excluding savannas and woodlands in which trees have a lower canopy cover. Other definitions consider savannas to be a type of forest, and include all areas with tree canopies over 10%.

Some areas covered in trees are legally defined as agricultural areas, e.g. Norway spruce plantations in Austrian forest law when the trees are being grown as Christmas trees and below a certain height.

Etymology

The word *forest* comes from Middle English, from Old French *forest* (also *forès*) "forest, vast expanse covered by trees"; first introduced in English as the word for wild land set aside for hunting without the necessity in definition for the existence of trees. Possibly a borrowing (probably via Frankish or Old High German) of the Medieval Latin word *foresta* "open wood", *foresta* was first used by Carolingian scribes in the Capitularies of Charlemagne to refer specifically to the king's royal hunting grounds. The term was not endemic to Romance languages (e.g. native words for "forest" in the Romance languages evolved out of the Latin word *silva* "forest, wood" (English *sylvan*); cf. Italian, Spanish, Portuguese *selva*; Romanian *silvă*; Old French *selve*); and cognates in Romance languages, such as Italian *foresta*, Spanish and Portuguese *floresta*, etc. are all ultimately borrowings of the French word.

Since the 13th century, the Niepołomice Forest in Poland has had special use and protection.
In this view from space, different coloration can indicate different functions.

The exact origin of Medieval Latin *foresta* is obscure. Some authorities claim the word derives from the Late Latin phrase *forestam silvam*, meaning "the outer wood"; others claim the term is a latinisation of the Frankish word **forhist* "forest, wooded country", assimilated to *forestam silvam* (a common practice among Frankish scribes). Frankish **forhist* is attested by Old High German *forst* "forest", Middle Low German *vorst* "forest", Old English *fyrhþ* "forest, woodland, game preserve, hunting ground" (English *frith*), and Old Norse *fýri* "coniferous forest", all of which derive from Proto-Germanic **furhisa-*, **furhíþija-* "a fir-wood, coniferous forest", from Proto-Indo-European **perkʷu-* "a coniferous or mountain forest, wooded height".

Uses of the word "forest" in English to denote any uninhabited area of non-enclosure are now considered archaic. The word was introduced by the Norman rulers of England as a legal term (appearing in Latin texts like the Magna Carta) denoting an uncultivated area legally set aside for hunting by feudal nobility.

Tywi Forest, Wales

These hunting forests were not necessarily wooded much, if at all. However, as hunting forests did often include considerable areas of woodland, the word "forest" eventually came to mean wooded land more generally. By the start of the 14th century, the word appeared in English texts, indicating all three senses: the most common one, the legal term and the archaic usage. Other terms used to mean "an area with a high density of trees" are *wood, woodland, wold, weald, holt, frith* and *firth*. Unlike *forest*, these are all derived from Old English and were not borrowed from another language. Some classifications now reserve the term *woodland* for an area with more open space between trees and distinguish among woodlands, *open forests*, and *closed forests* based on crown cover.

Evolution

The first known forests on Earth arose in the Late Devonian (approximately 380 million years ago), with the evolution of *Archaeopteris*. *Archaeopteris* was a plant that was both tree-like and fern-like, growing to 10 metres (33 ft) in height. *Archaeopteris* quickly spread throughout the world, from the equator to subpolar latitudes. *Archaeopteris* formed the first forest by being the first known species to cast shade due to its fronds and forming soil from its roots. *Archaeopteris* was deciduous, dropping its fronds onto the forest floor. The shade, soil, and forest duff from the dropped fronds created the first forest. The shed organic matter altered the freshwater environment, slowing it down and providing food. This promoted freshwater fish.

Ecology

Temperate rainforest in Tasmania's Hellyer Gorge

Forests account for 75% of the gross primary productivity of the Earth's biosphere, and contain 80% of the Earth's plant biomass. Forest ecosystems can be found in all regions capable of sustaining tree growth, at altitudes up to the tree line, except where natural fire frequency or other disturbance is too high, or where the environment has been altered by human activity.

The latitudes 10° north and south of the equator are mostly covered in tropical rainforest, and the latitudes between 53°N and 67°N have boreal forest. As a general rule, forests dominated by angiosperms (*broadleaf forests*) are more species-rich than those dominated by gymnosperms (*conifer, montane,* or *needleleaf forests*), although exceptions exist.

Forests sometimes contain many tree species within a small area (as in tropical rain and temperate deciduous forests), or relatively few species over large areas (e.g., taiga and arid montane coniferous forests). Forests are often home to many animal and plant species, and biomass per unit area is high compared to other vegetation communities. Much of this biomass occurs below ground in the root systems and as partially decomposed plant detritus. The woody component of a forest contains lignin, which is relatively slow to decompose compared with other organic materials such as cellulose or carbohydrate.

Components

Even, dense old-growth stand of beech trees (*Fagus sylvatica*) prepared to be regenerated by their saplings in the understory, in the Brussels part of the Sonian Forest.

A forest consists of many components that can be broadly divided into two categories that are biotic (living) and abiotic (non-living) components. The living parts include trees, shrubs, vines, grasses and other herbaceous (non-woody) plants, mosses, algae, fungi, insects, mammals, birds, reptiles, amphibians, and microorganisms living on the plants and animals and in the soil.

Layers

Biogradska forest in Montenegro

A forest is made up of many layers. Starting from the ground level and moving up, the main layers of all forest types are the forest floor, the understory and the canopy. The emergent layer exists in tropical rainforests. Each layer has a different set of plants and animals depending upon the availability of sunlight, moisture and food.

- Forest floor contains decomposing leaves, animal droppings, and dead trees. Decay on the forest floor forms new soil and provides nutrients to the plants. The forest floor supports ferns, grasses, mushroom and tree seedlings.

- Understory is made up of bushes, shrubs, and young trees that are adapted to living in the shades of the canopy.

- Canopy is formed by the mass of intertwined branches, twigs and leaves of the mature trees. The crowns of the dominant trees receive most of the sunlight. This is the most productive part of the trees where maximum food is produced. The canopy forms a shady, protective "umbrella" over the rest of the forest.

- Emergent layer exists in the tropical rain forest and is composed of a few scattered trees that tower over the canopy.

Spiny forest at Ifaty, Madagascar, featuring various *Adansonia* (baobab) species, *Alluaudia procera* (Madagascar ocotillo) and other vegetation

Types

Forests can be classified in different ways and to different degrees of specificity. One such way is in terms of the biome in which they exist, combined with leaf longevity of the dominant species (whether they are evergreen or deciduous). Another distinction is whether the forests are composed predominantly of broadleaf trees, coniferous (needle-leaved) trees, or mixed.

- Boreal forests occupy the subarctic zone and are generally evergreen and coniferous.

- Temperate zones support both broadleaf deciduous forests (*e.g.*, temperate deciduous forest) and evergreen coniferous forests (*e.g.*, temperate coniferous forests and temperate rainforests). Warm temperate zones support broadleaf evergreen forests, including laurel forests.

- Tropical and subtropical forests include tropical and subtropical moist forests, tropical and subtropical dry forests, and tropical and subtropical coniferous forests.

- Physiognomy classifies forests based on their overall physical structure or developmental stage (e.g. old growth vs. second growth).

- Forests can also be classified more specifically based on the climate and the dominant tree species present, resulting in numerous different forest types (e.g., Ponderosa pine/Douglas-fir forest).

The number of trees in the world, according to a 2015 estimate, is 3 trillion, of which 1.4 trillion are in the tropics or sub-tropics, 0.6 trillion in the temperate zones, and 0.7 trillion in the coniferous boreal forests. The estimate is about eight times higher than previous estimates, and is based on tree densities measured on over 400,000 plots. It remains subject to a wide margin of error, not least because the samples are mainly from Europe and North America.

A dry sclerophyll forest in Sydney, which is dominated by eucalyptus trees.

Forests can also be classified according to the amount of human alteration. Old-growth forest contains mainly natural patterns of biodiversity in established seral patterns, and they contain mainly species native to the region and habitat. In contrast, secondary forest is regrowing forest following timber harvest and may contain species originally from other regions or habitats.

Different global forest classification systems have been proposed, but none has gained universal acceptance. UNEP-WCMC's forest category classification system is a simplification of other more complex systems (e.g. UNESCO's forest and woodland 'subformations'). This system divides the world's forests into 26 major types, which reflect climatic zones as well as the principal types of trees. These 26 major types can be reclassified into 6 broader categories: temperate needleleaf; temperate broadleaf and mixed; tropical moist; tropical dry; sparse trees and parkland; and forest plantations.

Temperate Needleleaf

Temperate needleleaf forests mostly occupy the higher latitude regions of the Northern Hemisphere, as well as high altitude zones and some warm temperate areas, especially on nutrient-poor or otherwise unfavourable soils. These forests are composed entirely, or nearly so, of coniferous species (Coniferophyta). In the Northern Hemisphere pines *Pinus*, spruces *Picea*, larches *Larix*, firs *Abies*, Douglas firs *Pseudotsuga* and hemlocks *Tsuga*, make up the canopy, but other taxa are also important. In the Southern Hemisphere, most coniferous trees (members of the Araucariaceae and Podocarpaceae) occur in mixtures with broadleaf species, and are classed as broadleaf and mixed forests.

Temperate Broadleaf and Mixed

Temperate broadleaf and mixed forests include a substantial component of trees in the Anthophyta. They are generally characteristic of the warmer temperate latitudes, but extend to cool temperate ones, particularly in the southern hemisphere. They include such forest types as the mixed deciduous forests of the United States and their counterparts in China and Japan, the broadleaf evergreen rainforests of Japan, Chile and Tasmania, the sclerophyllous forests of Australia, central Chile, the Mediterranean and California, and the southern beech Nothofagus forests of Chile and New Zealand.

Broadleaf forest in Bhutan

Tropical Moist

There are many different types of tropical moist forests, with lowland evergreen broad leaf tropical rainforests, for example várzea and igapó forests and the terra firma forests of the Amazon Basin; the peat swamp forests, dipterocarp forests of Southeast Asia; and the high forests of the Congo Basin. Seasonal tropical forests, perhaps the best description for the colloquial term "jungle", typically range from the rainforest zone 10 degrees north or south of the equator, to the Tropic of Cancer and Tropic of Capricorn. Forests located on mountains are also included in this category, divided largely into upper and lower montane formations on the basis of the variation of physiognomy corresponding to changes in altitude.

Tropical Dry

Tropical dry forests are characteristic of areas in the tropics affected by seasonal drought. The seasonality of rainfall is usually reflected in the deciduousness of the forest canopy, with most trees being leafless for several months of the year. However, under some conditions, e.g. less fertile soils or less predictable drought regimes, the proportion of evergreen species increases and the forests are characterised as "sclerophyllous". Thorn forest, a dense forest of low stature with a high frequency of thorny or spiny species, is found where drought is prolonged, and especially where grazing animals are plentiful. On very poor soils, and especially where fire or herbivory are recurrent phenomena, savannas develop.

Sparse Trees and Parkland

Sparse trees and savanna are forests with lower canopy cover of trees. They occur principally in areas of transition from forested to non-forested landscapes. The two major zones in which these ecosystems occur are in the boreal region and in the seasonally dry tropics. At high latitudes, north of the main zone of boreal forest, growing conditions are not adequate to maintain a continuous closed forest cover, so tree cover is both sparse and discontinuous. This vegetation is variously called open taiga, open lichen woodland, and forest tundra. A savanna is a mixed woodland grass-land ecosystem characterized by the trees being sufficiently widely spaced so that the canopy does not close. The open canopy allows sufficient light to reach the ground to support an unbroken her-baceous layer consisting primarily of grasses. Savannas maintain an open canopy despite a high tree density.

Taiga forest near Saranpaul in the northeast Ural Mountains, Khanty–Mansia, Russia. Trees include *Picea obovata* (dominant on right bank), *Larix sibirica*, *Pinus sibirica*, and *Betula pendula*.

Forest Plantations

Forest plantations are generally intended for the production of timber and pulpwood. Commonly mono-specific and/or composed of introduced tree species, these ecosystems are not generally important as habitat for native biodiversity. However, they can be managed in ways that enhance their biodiversity protection functions and they can provide ecosystem services such as maintain-ing nutrient capital, protecting watersheds and soil structure, and storing carbon.

Societal Significance

Coastal Douglas fir woodland in northwest Oregon

Forests provide a diversity of ecosystem services including converting carbon dioxide into oxygen and biomass, acting as a carbon sink, aiding in regulating climate, purifying water, mitigating natural hazards such as floods, and serving as a genetic reserve. Forests also serve as a source of lumber and as recreational areas.

Redwood tree in northern California redwood forest, where many redwood trees are managed for preservation and longevity, rather than being harvested for wood production

Some researchers state that forests do not only provide benefits, but can in certain cases also incur costs to humans. Forests may impose an economic burden, diminish the enjoyment of natural areas, reduce the food producing capacity of grazing land and cultivated land, reduce biodiversity reduce available water for humans and wildlife, harbour dangerous or destructive wildlife, and act as reservoirs of human and livestock disease.

The management of forests is often referred to as forestry. Forest management has changed considerably over the last few centuries, with rapid changes from the 1980s onwards culminating in a practice now referred to as sustainable forest management. Forest ecologists concentrate on forest patterns and processes, usually with the aim of elucidating cause-and-effect relationships. Foresters who practice sustainable forest management focus on the integration of ecological, social, and economic values, often in consultation with local communities and other stakeholders.

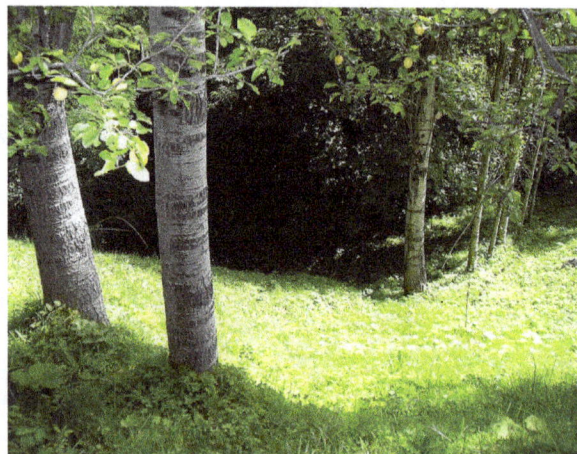

A forest near Vinitsa, Republic of Macedonia

Humans have generally decreased the amount of forest worldwide. Anthropogenic factors that can affect forests include logging, urban sprawl, human-caused forest fires, acid rain, invasive species, and the slash and burn practices of swidden agriculture or shifting cultivation. The loss and re-growth of forest leads to a distinction between two broad types of forest, primary or old-growth forest and secondary forest. There are also many natural factors that can cause changes in forests over time including forest fires, insects, diseases, weather, competition between species, etc. In 1997, the World Resources Institute recorded that only 20% of the world's original forests remained in large intact tracts of undisturbed forest. More than 75% of these intact forests lie in three countries—the boreal forests of Russia and Canada and the rainforest of Brazil.

In 2010, the Food and Agriculture Organization of the United Nations reported that world deforestation, mainly the conversion of tropical forests to agricultural land, had decreased over the past ten years but still continues at a high rate in many countries. Globally, around 13 million hectares of forests were converted to other uses or lost through natural causes each year between 2000 and 2010 as compared to around 16 million hectares per year during the 1990s. The study covered 233 countries and areas. Brazil and Indonesia, which had the highest loss of forests in the 1990s, have significantly reduced their deforestation rates. China instituted a ban on logging, beginning in 1998, due to the erosion and flooding that it caused. In addition, ambitious tree planting programmes in countries such as China, India, the United States and Vietnam - combined with natural expansion of forests in some regions - have added more than seven million hectares of new forests annually. As a result, the net loss of forest area was reduced to 5.2 million hectares per year between 2000 and 2010, down from 8.3 million hectares annually in the 1990s. In 2015, a study for *Nature Climate Change* showed that the trend has recently been reversed, leading to an "overall gain" in global biomass and forests. This gain is due especially to reforestation in China and Russia. However new forests are not completely equivalent to old growth forests in terms of species diversity, resilience and carbon capture. On September 7, 2015, the Food and Agriculture Organization of the United Nations released a new study stating that, over the last 25 years, the global deforestation rate has decreased by 50% due to improved management of forests and greater government protection.

Smaller areas of woodland in cities may be managed as urban forestry, sometimes within public parks. These are often created for human benefits; Attention Restoration Theory argues that spending time in nature reduces stress and improves health, while forest schools and kindergartens help young people to develop social as well as scientific skills in forests. These typically need to be close to where the children live, for practical logistics.

Canada

Garibaldi Provincial Park, British Columbia

Canada has about 4,020,000 square kilometres (1,550,000 sq mi) of forest land. More than 90% of forest land is publicly owned and about 50% of the total forest area is allocated for harvesting. These allocated areas are managed using the principles of sustainable forest management, which includes extensive consultation with local stakeholders. About eight percent of Canada's forest is legally protected from resource development. Much more forest land—about 40 percent of the total forest land base—is subject to varying degrees of protection through processes such as integrated land use planning or defined management areas such as certified forests.

By December 2006, over 1,237,000 square kilometers of forest land in Canada (about half the global total) had been certified as being sustainably managed. Clearcutting, first used in the latter half of the 20th century, is less expensive, but devastating to the environment, and companies are required by law to ensure that harvested areas are adequately regenerated. Most Canadian provinces have regulations limiting the size of clearcuts, although some older clearcuts can range upwards of 110 square kilometres (27,000 acres) in size which were cut over several years.

United States

Priest River winding through Whitetail Butte with lots of forestry to the east—these lot patterns have existed since the mid-19th century. The white patches reflect areas with younger, smaller trees, where winter snow cover shows up brightly to the astronauts. Dark green-brown squares are parcels of denser, intact forest.

In the United States, most forests have historically been affected by humans to some degree, though in recent years improved forestry practices has helped regulate or moderate large scale or severe impacts. However, the United States Forest Service estimates a net loss of about 2 million hectares (4,942,000 acres) between 1997 and 2020; this estimate includes conversion of forest land to other uses, including urban and suburban development, as well as afforestation and natural reversion of abandoned crop and pasture land to forest. However, in many areas of the United States, the area of forest is stable or increasing, particularly in many northern states. The opposite problem from flooding has plagued national forests, with loggers complaining that a lack of thinning and proper forest management has resulted in large forest fires.

Forestry

Forestry is the science and craft of creating, managing, using, conserving, and repairing forests and associated resources to meet desired goals, needs, and values for human and environment benefits. Forestry is practiced in plantations and natural stands. The science of forestry has elements that belong to the biological, physical, social, political and managerial sciences.

Forestry work in Austria

Modern forestry generally embraces a broad range of concerns, in what is known as multiple-use management, including the provision of timber, fuel wood, wildlife habitat, natural water quality management, recreation, landscape and community protection, employment, aesthetically appealing landscapes, biodiversity management, watershed management, erosion control, and preserving forests as 'sinks' for atmospheric carbon dioxide. A practitioner of forestry is known as a forester. Other terms are used a verderer and a silviculturalist being common ones. Silviculture is narrower than forestry, being concerned only with forest plants, but is often used synonymously with forestry.

Forest ecosystems have come to be seen as the most important component of the biosphere, and forestry has emerged as a vital applied science, craft, and technology.

Forestry is an important economic segment in various industrial countries. For example, in Germany, forests cover nearly a third of the land area, wood is the most important renewable resource, and forestry supports more than a million jobs and about billion in yearly turnover.

A deciduous beech forest in Slovenia

History

Background

The preindustrial age has been dubbed by Werner Sombart and others as the 'wooden age', as timber and firewood were the basic resources for energy, construction and housing. The development of modern forestry is closely connected with the rise of capitalism, economy as a science and varying notions of land use and property.

Roman Latifundiae, large agricultural estates, were quite successful in maintaining the large supply of wood that was necessary for the Roman Empire. Large deforestations came with respectively after the decline of the Romans. However already in the 5th century, monks in the then Byzantine Romagna on the Adriatic coast, were able to establish stone pine plantations to provide fuelwood and food. This was the beginning of the massive forest mentioned by Dante Alighieri in his 1308 poem Divine Comedy.

Similar sustainable formal forestry practices were developed by the Visigoths in the 7th century when, faced with the ever-increasing shortage of wood, they instituted a code concerned with the preservation of oak and pine forests. The use and management of many forest resources has a long history in China as well, dating back to the Han Dynasty and taking place under the landowning gentry. A similar approach was used in Japan. It was also later written about by the Ming Dynasty Chinese scholar Xu Guangqi (1562–1633).

In Europe, land usage rights in medieval and early modern times allowed different users to access forests and pastures. Plant litter and resin extraction were important, as pitch (resin) was essential for the caulking of ships, falking and hunting rights, firewood and building, timber gathering in wood pastures, and for grazing animals in forests. The notion of "commons" (German "Allmende") refers to the underlying traditional legal term of common land. The idea of enclosed private property came about during modern times. However, most hunting rights were retained by members of the nobility which preserved the right of the nobility to access and use common land for recreation, like fox hunting.

Early Modern Forestry Development

Timber harvesting, as here in Finland, is a common component of forestry

Hans Carl von Carlowitz

Systematic management of forests for a sustainable yield of timber is said to have begun in the German states in the 14th century, e.g. in Nuremberg, and in 16th-century Japan. Typically, a forest was divided into specific sections and mapped; the harvest of timber was planned with an eye to regeneration. As timber rafting allowed for connecting large continental forests, as in south western Germany, via Main, Neckar, Danube and Rhine with the coastal cities and states, early modern forestry and remote trading were closely connected. Large firs in the black forest were called „Holländer", as they were traded to the Dutch ship yards. Large timber rafts on the Rhine were 200 to 400m in length, 40m in width and consisted of several thousand logs. The crew consisted of 400 to 500 men, including shelter, bakeries, ovens and livestock stables. Timber rafting infrastructure allowed for large interconnected networks all over continental Europe and is still of importance in Finland.

Starting with the sixteenth century, enhanced world maritime trade, a boom in housing construction in Europe and the success and further Berggeschrey (rushes) of the mining industry increased timber consumption sharply. The notion of 'Nachhaltigkeit', sustainability in forestry, is closely connected to the work of Hans Carl von Carlowitz (1645–1714), a mining administrator in Saxony. His book *Sylvicultura oeconomica, oder haußwirthliche Nachricht und Naturmäßige Anweisung zur wilden Baum-Zucht* (1713) was the first comprehensive treatise about sustainable yield forestry. In the UK, and to an extend in continental Europe, the enclosure movement and the clearances favored strictly enclosed private property. The Agrarian reformers, early economic writers and scientists tried to get rid of the traditional commons. At the time, an alleged tragedy of the commons together with fears of a Holznot, an imminent wood shortage played a watershed role in the controversies about cooperative land use patterns.

The practice of establishing tree plantations in the British Isles was promoted by John Evelyn, though it had already acquired some popularity. Louis XIV's minister Jean-Baptiste Colbert's oak Forest of Tronçais, planted for the future use of the French Navy, matured as expected in the mid-19th century: "Colbert had thought of everything except the steamship," Fernand Braudel observed. In parallel, schools of forestry were established beginning in the late 18th century in Hesse, Russia, Austria-Hungary, Sweden, France and elsewhere in Europe.

Forest Conservation and Early Globalization

During the late 19th and early 20th centuries, forest preservation programs were established in British India, the United States, and Europe. Many foresters were either from continental Europe (like Sir Dietrich Brandis), or educated there (like Gifford Pinchot). Sir Dietrich Brandis is considered the father of tropical forestry, European concepts and practices had to be adapted in tropical and semi arid climate zones. The development of plantation forestry was one of the (controversial) answers to the specific challenges in the tropical colonies. The enactment and evolution of forest laws and binding regulations occurred in most Western nations in the 20th century in response to growing conservation concerns and the increasing technological capacity of logging companies. Tropical forestry is a separate branch of forestry which deals mainly with equatorial forests that yield woods such as teak and mahogany.

Mechanization

Forestry mechanization was always in close connection to metal working and the development of mechanical tools to cut and transport timber to its destination. Rafting belongs to the earliest means of transport. Steel saws came up in the 15th century. The 19th century widely increased the availability of steel for whipsaws and introduced Forest railways and railways in general for transport and as forestry customer. Further human induced changes, however, came since World War II, respectively in line with the '1950s-syndrome'. The first portable chainsaw was invented in 1918 in Canada, but large impact of mechanization in forestry started after World War II. Forestry harvesters are among the most recent developments. Although drones, planes, laser scanning, satellites and robots also play a part in forestry.

Forestry Today

Today a strong body of research exists regarding the management of forest ecosystems and genetic improvement of tree species and varieties. Forestry also includes the development of better methods for the planting, protecting, thinning, controlled burning, felling, extracting, and processing of timber. One of the applications of modern forestry is reforestation, in which trees are planted and tended in a given area.

A modern sawmill

Trees provide numerous environmental, social and economic benefits for people. In many regions the forest industry is of major ecological, economic, and social importance. Third-party certification systems that provide independent verification of sound forest stewardship and sustainable forestry have become commonplace in many areas since the 1990s. These certification systems were developed as a response to criticism of some forestry practices, particularly deforestation in less developed regions along with concerns over resource management in the developed world. Some certification systems are criticized for primarily acting as marketing tools and lacking in their claimed independence.

In topographically severe forested terrain, proper forestry is important for the prevention or minimization of serious soil erosion or even landslides. In areas with a high potential for landslides, forests can stabilize soils and prevent property damage or loss, human injury, or loss of life.

Public perception of forest management has become controversial, with growing public concern over perceived mismanagement of the forest and increasing demands that forest land be managed for uses other than pure timber production, for example, indigenous rights, recreation, watershed management, and preservation of wilderness, waterways and wildlife habitat. Sharp disagreements over the role of forest fires, logging, motorized recreation and other issues drives debate while the public demand for wood products continues to increase.

Foresters

Foresters work for the timber industry, government agencies, conservation groups, local authorities, urban parks boards, citizens' associations, and private landowners. The forestry profession includes a wide diversity of jobs, with educational requirements ranging from college bachelor's degrees to PhDs for highly specialized work. Industrial foresters plan forest regeneration starting with careful harvesting. Urban foresters manage trees in urban green spaces. Foresters work in tree nurseries growing seedlings for woodland creation or regeneration projects. Foresters improve tree genetics. Forest engineers develop new building systems. Professional foresters measure and model the growth of forests with tools like geographic information systems. Foresters may combat insect infestation, disease, forest and grassland wildfire, but increasingly allow these natural aspects of forest ecosystems to run their course when the likelihood of epidemics or risk of life or property are low. Increasingly, foresters participate in wildlife conservation planning and watershed protection. Foresters have been mainly concerned with timber management, especially reforestation, maintaining forests at prime conditions, and fire control.

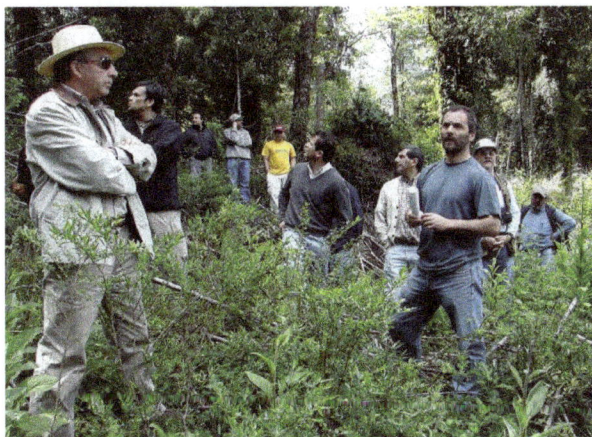

Foresters of UACh in the Valdivian forests of San Pablo de Tregua, Chile

Forestry Plans

Foresters develop and implement forest management plans relying on mapped resource inventories showing an area's topographical features as well as its distribution of trees (by species) and other plant cover. Plans also include landowner objectives, roads, culverts, proximity to human habitation, water features and hydrological conditions, and soils information. Forest management plans typically include recommended silvicultural treatments and a timetable for their implementation. Application of digital maps in Geographic Informations systems (GIS) that extracts and integrates different information about forest terrains, soil type and tree covers, etc. using, e.g. laser scanning, enhances forest management plans in modern systems.

Forest management plans include recommendations to achieve the landowner's objectives and desired future condition for the property subject to ecological, financial, logistical (e.g. access to resources), and other constraints. On some properties, plans focus on producing quality wood products for processing or sale. Hence, tree species, quantity, and form, all central to the value of harvested products quality and quantity, tend to be important components of silvicultural plans.

Good management plans include consideration of future conditions of the stand after any recommended harvests treatments, including future treatments (particularly in intermediate stand treatments), and plans for natural or artificial regeneration after final harvests.

The objectives of landowners and leaseholders influence plans for harvest and subsequent site treatment. In Britain, plans featuring "good forestry practice" must always consider the needs of other stakeholders such as nearby communities or rural residents living within or adjacent to woodland areas. Foresters consider tree felling and environmental legislation when developing plans. Plans instruct the sustainable harvesting and replacement of trees. They indicate whether road building or other forest engineering operations are required.

Agriculture and forest leaders are also trying to understand how the climate change legislation will affect what they do. The information gathered will provide the data that will determine the role of agriculture and forestry in a new climate change regulatory system.

Forestry as a Science

Over the past centuries, forestry was regarded as a separate science. With the rise of ecology and environmental science, there has been a reordering in the applied sciences. In line with this view, forestry is one of three primary land-use sciences. The other two are agriculture and agroforestry. Under these headings, the fundamentals behind the management of natural forests comes by way of natural ecology. Forests or tree plantations, those whose primary purpose is the extraction of forest products, are planned and managed utilizing a mix of ecological and agroecological principles.

Genetic Diversity in Forestry

The provenance of forest reproductive material used to plant forests has great influence on how the trees develop why it is important to use forest reproductive material of good quality and of high genetic diversity

The term, *genetic diversity* describe differences in DNA sequence between individuals as distinct from variation caused by environmental influences. The unique genetic composition of an individual (its genotype) will determine its performance (its phenotype) at a particular site.

Genetic diversity is needed to maintain the vitality of forests and to provide resilience to pests and diseases. Genetic diversity also ensures that forest trees can survive, adapt and evolve under changing environmental conditions. Furthermore, genetic diversity is the foundation of biological diversity at species and ecosystem levels. Forest genetic resources are therefore important to consider in forest management.

Genetic diversity in forests is threatened by forest fires, pests and diseases, habitat fragmentation, poor silvicultural practices and inappropriate use of forest reproductive material. Furthermore, the marginal populations of many tree species are facing new threats due to climate change.

Most countries in Europe have recommendations or guidelines for selecting species and provenances that can be used in a given site or zone.

Education

History of Forestry Education

The first dedicated forestry school was established by Georg Ludwig Hartig at Hungen in the Wetterau, Hesse, in 1787, though forestry had been taught earlier in central Europe, including at the University of Giessen, in Hesse-Darmstadt.

In Spain, the first forestry school was the Forest Engineering School of Madrid (Escuela Técnica Superior de Ingenieros de Montes), founded in 1844.

The first in North America, the Biltmore Forest School was established near Asheville, North Carolina, by Carl A. Schenck on September 1, 1898, on the grounds of George W. Vanderbilt's Biltmore Estate. Another early school was the New York State College of Forestry, established at Cornell University just a few weeks later, in September 1898. Early 19th century North American foresters went to Germany to study forestry. Some early German foresters also emigrated to North America.

Forestry Education Today

Prescribed burning is used by foresters to reduce fuel loads

Today, forestry education typically includes training in general biology, botany, genetics, soil science, climatology, hydrology, economics and forest management. Education in the basics of sociology and political science is often considered an advantage.

In India, forestry education is imparted in the agricultural universities and in Forest Research Institutes (deemed universities). Four year degree programmes are conducted in these universities at the undergraduate level. Masters and Doctorate degrees are also available in these universities.

In the United States, postsecondary forestry education leading to a Bachelor's degree or Master's degree is accredited by the Society of American Foresters.

In Canada the Canadian Institute of Forestry awards silver rings to graduates from accredited university BSc programs, as well as college and technical programs.

In many European countries, training in forestry is made in accordance with requirements of the Bologna Process and the European Higher Education Area.

The International Union of Forest Research Organizations is the only international organization that coordinates forest science efforts worldwide.

Sustainable Forest Management

Sustainable forest management is the management of forests according to the principles of sustainable development. Sustainable forest management has to keep the balance between three main pillars: ecological, economic and socio-cultural. Successfully achieving sustainable forest management will provide integrated benefits to all, ranging from safeguarding local livelihoods to protecting the biodiversity and ecosystems provided by forests, reducing rural poverty and mitigating some of the effects of climate change.

The "Forest Principles" adopted at The United Nations Conference on Environment and Development (UNCED) in Rio de Janeiro in 1992 captured the general international understanding of sustainable forest management at that time. A number of sets of criteria and indicators have since been developed to evaluate the achievement of SFM at the global, regional, country and management unit level. These were all attempts to codify and provide for independent assessment of the degree to which the broader objectives of sustainable forest management are being achieved in practice. In 2007, the United Nations General Assembly adopted the Non-Legally Binding Instrument on All Types of Forests. The instrument was the first of its kind, and reflected the strong international commitment to promote implementation of sustainable forest management through a new approach that brings all stakeholders together.

Definition

A definition of SFM was developed by the Ministerial Conference on the Protection of Forests in Europe (FOREST EUROPE), and has since been adopted by the Food and Agriculture Organization (FAO). It defines sustainable forest management as:

The stewardship and use of forests and forest lands in a way, and at a rate, that maintains their biodiversity, productivity, regeneration capacity, vitality and their potential to fulfill, now and in the future, relevant ecological, economic and social functions, at local, national, and global levels, and that does not cause damage to other ecosystems.

In simpler terms, the concept can be described as the attainment of balance – balance between society's increasing demands for forest products and benefits, and the preservation of forest health and diversity. This balance is critical to the survival of forests, and to the prosperity of forest-dependent communities.

For forest managers, sustainably managing a particular forest tract means determining, in a tangible way, how to use it today to ensure similar benefits, health and productivity in the future. Forest managers must assess and integrate a wide array of sometimes conflicting factors – commercial and non-commercial values, environmental considerations, community needs, even global impact – to produce sound forest plans. In most cases, forest managers develop their forest plans in consultation with citizens, businesses, organizations and other interested parties in and around the forest tract being managed. The tools and visualization have been recently evolving for better management practices.

The Food and Agriculture Organization of the United Nations, at the request of Member States, developed and launched the Sustainable Forest Management Toolbox in 2014, an online collection of tools, best practices and examples of their application to support countries implementing sustainable forest management.

Because forests and societies are in constant flux, the desired outcome of sustainable forest management is not a fixed one. What constitutes a sustainably managed forest will change over time as values held by the public change.

Criteria and Indicators

Deforestation of native rain forest in Rio de Janeiro City for extraction of clay for civil engineering (2009 picture). An example of non sustainable forest management.

Criteria and indicators are tools which can be used to conceptualise, evaluate and implement sustainable forest management. Criteria define and characterize the essential elements, as well as a set of conditions or processes, by which sustainable forest management may be assessed. Periodically measured indicators reveal the direction of change with respect to each criterion.

Criteria and indicators of sustainable forest management are widely used and many countries produce national reports that assess their progress toward sustainable forest management. There are nine international and regional criteria and indicators initiatives, which collectively involve more than 150 countries. Three of the more advanced initiatives are those of the Working Group on Criteria and Indicators for the Conservation and Sustainable Management of Temperate and Boreal Forests (also called the Montreal Process), Forest Europe, and the International Tropical Timber Organization. Countries who are members of the same initiative usually agree to produce reports at the same time and using the same indicators. Within countries, at the management unit level, efforts have also been directed at developing local level criteria and indicators of sustainable forest management. The Center for International Forestry Research, the International Model Forest Network and researchers at the University of British Columbia have developed a number of tools and techniques to help forest-dependent communities develop their own local level criteria and indicators. Criteria and Indicators also form the basis of third-party forest certification programs such as the Canadian Standards Association's Sustainable Forest Management Standards and the Sustainable Forestry Initiative Standard.

There appears to be growing international consensus on the key elements of sustainable forest management. Seven common thematic areas of sustainable forest management have emerged based on the criteria of the nine ongoing regional and international criteria and indicators initiatives. The seven thematic areas are:

- Extent of forest resources

- Biological diversity

- Forest health and vitality

- Productive functions and forest resources

- Protective functions of forest resources

- Socio-economic functions

- Legal, policy and institutional framework.

This consensus on common thematic areas (or criteria) effectively provides a common, implicit definition of sustainable forest management. The seven thematic areas were acknowledged by the international forest community at the fourth session of the United Nations Forum on Forests and the 16th session of the Committee on Forestry. These thematic areas have since been enshrined in the Non-Legally Binding Instrument on All Types of Forests as a reference framework for sustainable forest management to help achieve the purpose of the instrument.

On January 5, 2012, the Montreal Process, Forest Europe, the International Tropical Timber Organization, and the Food and Agriculture Organization of the United Nations, acknowledging the seven thematic areas, endorsed a joint statement of collaboration to improve global forest related data collection and reporting and avoiding the proliferation of monitoring requirements and associated reporting burdens.

Ecosystem Approach

The Ecosystem Approach has been prominent on the agenda of the Convention on Biological Diversity (CBD) since 1995 . The CBD definition of the Ecosystem Approach and a set of principles for its application were developed at an expert meeting in Malawi in 1995, known as the Malawi Principles. The definition, 12 principles and 5 points of "operational guidance" were adopted by the fifth Conference of Parties (COP5) in 2000. The CBD definition is as follows

The ecosystem approach is a strategy for the integrated management of land, water and living resources that promotes conservation and sustainable use in an equitable way. Application of the ecosystem approach will help to reach a balance of the three objectives of the Convention. An ecosystem approach is based on the application of appropriate scientific methodologies focused on levels of biological organization, which encompasses the essential structures, processes, functions and interactions among organisms and their environment. It recognizes that humans, with their cultural diversity, are an integral component of many ecosystems.

Sustainable forest management was recognized by parties to the Convention on Biological Diversity in 2004 (Decision VII/11 of COP7) to be a concrete means of applying the Ecosystem Approach to forest ecosystems. The two concepts, sustainable forest management and the ecosystem approach, aim at promoting conservation and management practices which are environmentally, socially and economically sustainable, and which generate and maintain benefits for both present and future generations. In Europe, the MCPFE and the Council for the Pan-European Biological and Landscape Diversity Strategy (PEBLDS) jointly recognized sustainable forest management to be consistent with the Ecosystem Approach in 2006.

Independent Certification

Growing environmental awareness and consumer demand for more socially responsible businesses helped third-party forest certification emerge in the 1990s as a credible tool for communicating the environmental and social performance of forest operations.

There are many potential users of certification, including: forest managers, scientists, policy makers, investors, environmental advocates, business consumers of wood and paper, and individuals.

With third-party forest certification, an independent organization develops standards of good forest management, and independent auditors issue certificates to forest operations that comply with those standards. Forest certification verifies that forests are well-managed – as defined by a particular standard – and chain-of-custody certification tracks wood and paper products from the certified forest through processing to the point of sale.

This rise of certification led to the emergence of several different systems throughout the world. As a result, there is no single accepted forest management standard worldwide, and each system takes a somewhat different approach in defining standards for sustainable forest management.

In its 2009–2010 Forest Products Annual Market Review United Nations Economic Commission for Europe/Food and Agriculture Organization stated: "Over the years, many of the issues that previously divided the (certification) systems have become much less distinct. The largest certification systems now generally have the same structural programmatic requirements."

Third-party forest certification is an important tool for those seeking to ensure that the paper and wood products they purchase and use come from forests that are well-managed and legally harvested. Incorporating third-party certification into forest product procurement practices can be a centerpiece for comprehensive wood and paper policies that include factors such as the protection of sensitive forest values, thoughtful material selection and efficient use of products.

There are more than fifty certification standards worldwide, addressing the diversity of forest types and tenures. Globally, the two largest umbrella certification programs are:

- Programme for the Endorsement of Forest Certification (PEFC)
- Forest Stewardship Council (FSC)

The area of forest certified worldwide is growing slowly. PEFC is the world's largest forest certification system, with more than two-thirds of the total global certified area certified to its Sustainability Benchmarks.

In North America, there are three certification standards endorsed by PEFC – the Sustainable Forestry Initiative, the Canadian Standards Association's Sustainable Forest Management Standard, and the American Tree Farm System. FSC has five standards in North America – one in the United States and four in Canada.

While certification is intended as a tool to enhance forest management practices throughout the world, to date most certified forestry operations are located in Europe and North America. A significant barrier for many forest managers in developing countries is that they lack the capacity to undergo a certification audit and maintain operations to a certification standard.

Forest Governance

Although a majority of forests continue to be owned formally by government, the effectiveness of forest governance is increasingly independent of formal ownership. Since neo-liberal ideology in the 1980s and the emanation of the climate change challenges, evidence that the state is failing to effectively manage environmental resources has emerged. Under neo-liberal regimes in the developing countries, the role of the state has diminished and the market forces have increasingly taken over the dominant socio-economic role. Though the critiques of neo-liberal policies have maintained that market forces are not only inappropriate for sustaining the environment, but are in fact a major cause of environmental destruction. Hardin's tragedy of the common (1968) has shown that the people cannot be left to do as they wish with land or environmental resources. Thus, decentralization of management offers an alternative solution to forest governance.

The shifting of natural resource management responsibilities from central to state and local governments, where this is occurring, is usually a part of broader decentralization process. According to Rondinelli and Cheema (1983), there are four distinct decentralization options: these are: (i) Privatization – the transfer of authority from the central government to non-governmental sectors otherwise known as market-based service provision, (ii) Delegation – centrally nominated local authority, (iii) Devolution – transfer of power to locally acceptable authority and (iv) Deconcentration – the redistribution of authority from the central government to field delegations of the central government. The major key to effective decentralization is increased broad-based participation in local-public decision making. In 2000, the World Bank report reveals that local government knows the needs and desires of their constituents better than the national government, while at the same time, it is easier to hold local leaders accountable. From the study of West African tropical forest, it is argued that the downwardly accountable and/or representative authorities with meaningful discretional powers are the basic institutional element of decentralization that should lead to efficiency, development and equity. This collaborates with the World Bank report in 2000 which says that decentralization should improve resource allocation, efficiency, accountability and equity "by linking the cost and benefit of local services more closely".

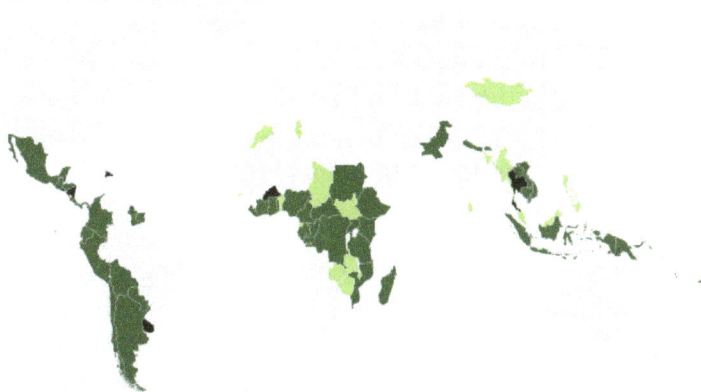

Countries participating in the UNREDD program and/or Forest Carbon Partnership Facility.

UN-REDD participants
Forest Carbon Partnership Facility participants
participants in both

Many reasons point to the advocacy of decentralization of forest. (i) Integrated rural development projects often fail because they are top-down projects that did not take local people's needs and desires into account. (ii) National government sometimes have legal authority over vast forest areas that they cannot control, thus, many protected area projects result in increased biodiversity loss and greater social conflict. Within the sphere of forest management, as state earlier, the most effective option of decentralization is "devolution"-the transfer of power to locally accountable authority. However, apprehension about local governments is not unfounded. They are often short of resources, may be staffed by people with low education and are sometimes captured by local elites who promote clientelist relation rather than democratic participation. Enters and Anderson (1999) point that the result of community-based projects intended to reverse the problems of past central approaches to conservation and development have also been discouraging.

Broadly speaking, the goal of forest conservation has historically not been met when, in contrast with land use changes; driven by demand for food, fuel and profit. It is necessary to recognize and advocate for better forest governance more strongly given the importance of forest in meeting basic human needs in the future and maintaining ecosystem and biodiversity as well as addressing climate change mitigation and adaptation goal. Such advocacy must be coupled with financial incentives for government of developing countries and greater governance role for local government, civil society, private sector and NGOs on behalf of the "communities".

National Forest Funds

The development of National Forest Funds is one way to address the issue of financing sustainable forest management. National forest funds (NFFs) are dedicated financing mechanisms managed by public institutions designed to support the conservation and sustainable use of forest resources. As of 2014, there are 70 NFFs operating globally.

Forest Genetic Resources

Appropriate use and long-term conservation of forest genetic resources (FGR) is a part of sustainable forest management. In particular when it comes to the adaptation of forests and forest management to climate change. Genetic diversity ensures that forest trees can survive, adapt and evolve under changing environmental conditions. Genetic diversity in forests also contributes to tree vitality and to the resilience towards pests and diseases. Furthermore, FGR has a crucial role in maintaining forest biological diversity at both species and ecosystem levels.

Selecting carefully the plant material with emphasis on getting a high genetic diversity rather than aiming at producing a uniform stand of trees, is essential for sustainable use of FGR. Considering the provenance is crucial as well. For example in relation to climate change, local material may not have the genetic diversity or phenotypic plasticity to guarantee good performance under changed conditions. A different population from further away, which may have experienced selection under conditions more like those forecast for the site to be reforested, might represent a more suitable seed source.

Criteria and Indicators of Sustainable Forest Management

Criteria & Indicators of Sustainable Forest Management (C&I) are policy instruments by which sustainability of forest management in the country/region, or progress towards Sustainable forest management (SFM), may be evaluated and reported on. C&I is a conjunctive term for a set of objectives and the variables/descriptions allowing to evaluate whether the objectives are achieved or not.

There are many various sets of C&I in the world that are used by particular regional SFM processes (e.g. FOREST EUROPE, Montreal Process), international organisations and their activities (e.g. FAO Global Forest Resources Assessment) or certification of forest management and forest products (e.g. Forest Stewardship Council, Programme for the Endorsement of Forest Certification). Signatory countries of particular processes or certification schemes can develop their national sets derived from the set of process/scheme.

Possible uses of Criteria and Indicators

Criteria and indicators of sustainable forest management:

- help to define, understand and promote the concept of sustainable forest management;
- provide a common framework for signatory countries to:
  - o describe, monitor, assess and report on national forest trends (if measured periodically);
  - o describe, monitor, assess and report on progress towards sustainable forest management (if measured periodically);

- reflect a holistic approach to forests as ecosystems, highlighting the full range of forest values;

- facilitate policy dialogue and the development of policies or strategies ;

- help to implement forest related policies, plans and programmes;

- contribute to cross-sectoral sustainability assessments, as well as assessments for other sectors (e.g. environment, energy, climate change, agriculture, sustainable land management);

- guide forest management practice;

- help to identify the changes in forest management;

- help to develop forest certification principles, standards and indicators.

Levels of Use

Criteria and indicators are applied at different levels:

- global - the sets of C&I for global use have to be universal enough for the various biogeographic regions, countries at different stages of development of forestry (with different ability to provide data), etc;

- regional (supranational regions, such as pan-European, are meant);

- national and sub-national;

- local (forest management units, community forestry land, concession areas and other) - as being foccused on much smaller scale, the sets for this level can signifficantly differ from sets for the above-mentioned levels and they are used mainly for forest certification purposes.

Early History

The first FAO Global Forest Resources Assessment was published in 1948. It comprised five indicators, however, their main focus was to assess the availability of timber, not sustainability of forest management as a whole.

The history of the idea of C&I of SFM goes back to 1992 when the United Nations Conference on Environment and Development adopted the "Forest Principles"* and Chapter 11 of Agenda 21. At about the same time, the International Tropical Timber Organization (ITTO) did some pioneering work on "Criteria for the Measurement of Sustainable Tropical Forest Management." Following this summit, the concept of "criteria and indicators for sustainable forest management" gained increasing international attention.

Forest Europe: Pan-European criteria and indicators for sustainable forest management

The first set of Pan-European C&I, based on documents adopted by two Expert Level Follow-Up

Meetings of the Helsinki Conference, was adopted by the third Ministerial Conference on the Protection of Forests in Europe on 2–4 June 1998 in Lisbon/Portugal as an Annex 1 of the Resolution L2. This set consisted of 6 Criteria of SFM, 20 quantitative indicators and 80 descriptive indicators (4 per each quantitative indicator).

In the meantime, knowledge and data collection systems as well as information needs have gradually developed further. Thus, initiated through the Lisbon Conference, the MCPFE decided to improve the existing set. Document named Improved Pan-European Indicators for Sustainable Forest Management was adopted at expert level at the MCPFE Expert Level Meeting, 7–8 October 2002 in Vienna, Austria (before Vienna Conference). The 6 criteria remained unchanged, however, the number of quantitative indicators was increased to 35, and the system of descriptive indicators, renamed to qualitative indicators, was significantly simplified to 17 indicators.

The last update of Pan-European set was performed before Madrid Conference at the "Forest Europe" Expert Level Meeting 30 June – 2 July 2015, in Madrid, Spain. The set of quantitative indicators was slightly altered (34 indicators); the system of qualitative indicators was further simplified to 11 indicators, 5 of them now forming something like an "unofficial 7th criterion", while the remaining 6 are the official 6 criteria.

Criteria

Criteria characterise or define the essential elements or set of conditions or processes by which sustainable forest management may be assessed (MCPFE, 1998b). There are 6 criteria in the Pan-European set:

1. Maintenance and Appropriate Enhancement of Forest Resources and their Contribution to Global Carbon Cycles

2. Maintenance of Forest Ecosystem Health and Vitality

3. Maintenance and Encouragement of Productive Functions of Forests (Wood and Non-Wood)

4. Maintenance, Conservation and Appropriate Enhancement of Biological Diversity in Forest Ecosystems

5. Maintenance and Appropriate Enhancement of Protective Functions in Forest Management (notably soil and water)

6. Maintenance of other socioeconomic functions and conditions

Indicators

The indicators show changes over time for each criterion and demonstrate the progress made towards its specified objective (MCPFE, 1998a).

- Quantitative indicators are expressed in measurement units and the necessary date are collected via regular forest inventories, other field surveys, remote sensing, etc. Periodically measured indicators reveal the direction of change with respect to criterion. The list of quantitative indicators includes, for example, the forest area and growing stock (volume of living wood) for the Criterion 1, forest damage for the Criterion 2, increment and fellings for the Criterion 3, deadwood volume or naturalness classes for the Criterion 4, the area of protective forests for the Criterion 5, and contribution of forests to GDP or the area of recreation forests for the Criterion 6.

- Qualitative indicators have to be described and assessed and the data are collected using questionnaires. They are used to describe legal and institutional frameworks of forestry, as well as policies and instruments for the implementation of SFM.

Montreal Process

The Montreal Process Working Group on C&I for the Conservation and Sustainable Management of Temperate and Boreal Forests was launched in 1994 as a response to the Rio Forest Principles. Original set of C&I was adopted by Santiago Declaration in 1995. It consisted of 7 criteria and 67 indicators. Current set represents the 5th version of MP C&I and it has 7 criteria and 54 indicators (both qualitative and quantitative.

Criteria

Conservation of biological diversity

1. Maintenance of productive capacity of forest ecosystems

2. Maintenance of forest ecosystem health and vitality

3. Conservation and maintenance of soil and water resources

4. Maintenance of forest contribution to global carbon cycles

5. Maintenance and enhancement of long-term multiple socio- economic benefits to meet # the needs of societies

6. Legal, institutional and economic framework for forest conservation and sustainable management

References

- Bundeswaldinventur 2002, Bundesministerium für Ernährung, Landwirtschaft und Verbraucherschutz (BMELV), retrieved, 17 January 2010

- T. Mirov, Nicholas; Hasbrouck, Jean (1976). "6". The story of pines. Bloomington and London: Indiana University Press. p. 111. ISBN 0-253-35462-5

- The Nature of Mediterranean Europe: An Ecological History, by Alfred Thomas Grove, Oliver Rackham, Yale University Press, 2003, review at Yale university press Nature of Mediterranean Europe: An Ecological History (review) Brian M. Fagan, Journal of Interdisciplinary History, Volume 32, Number 3, Winter 2002, pp. 454-455

- "LEDS GP Agriculture, Forestry and Other Land Use Working Group factsheet" (PDF). Low Emission Development Strategies Global Partnership (LEDS GP). Retrieved 23 March 2016

- Braudel, Fernand (1979). The Wheels of Commerce: Civilization and Capitalism: 15th-18th Century (Volume II). University of California Press. p. 240. ISBN 978-0-520-08115-4

- Victor Giurgiu (Nov 2011). "Revista pădurilor (Journal of forests) 125 years of existence". Rev. pădur. 126 (6): 3–7. ISSN 1583-7890. Retrieved 2012-04-06

- "Sustainable Forest Management Toolbox" (PDF). Food and Agriculture Organization of the United Nations. Retrieved 24 June 2014

- Matta, Rao (2015). Towards effective national forest funds, FAO Forestry Paper 174 (PDF). Rome, Italy: Food and Agriculture Organization of the United Nations. ISBN 978-92-5-108706-0

- "Evans, K., De Jong, W., and Cronkleton, P. (2008) "Future Scenarios as a Tool for Collaboration in Forest Communities". "S.A.P.I.EN.S." "'1'" (2)". Sapiens.revues.org. 1 October 2008. Retrieved 30 November 2011

Afforestation and its Positive Impacts

Afforestation is the growing of trees in areas where there were previously no trees. Government organizations and non-profitable organizations engage in the practice of afforestation. This chapter is an overview of the subject matter incorporating all the major aspects of afforestation.

Afforestation

An afforestation project in Rand Wood, Lincolnshire, England

Afforestation is the establishment of a forest or stand of trees in an area where there was no previous tree cover. Reforestation is the reestablishment of forest cover, either naturally (by natural seeding, coppice, or root suckers) or artificially (by direct seeding or planting).

Forestation is the establishment of forest growth on areas that either had forest or lacked it. Reforestation and afforestation are categories of forestation. Many governments and non-governmental organizations directly engage in programs of *afforestation* to create forests, increase carbon capture and carbon sequestration, and help to anthropogenically improve biodiversity. (In the UK, afforestation may mean converting the legal status of some land to "royal forest".) Special tools, e.g. tree planting bar, are used to make planting of trees easier and faster.

Biological Process

Gap dynamics is the pattern of plant growth that occurs following the creation of a forest gap, a local area of natural disturbance that results in an opening in the canopy of a forest. Gap dynamics are a typical characteristic of temperate and tropical forests, and have a wide variety of causes and effects on forest life.

In Areas of Degraded Soil

In some places, forests need help to reestablish themselves because of environmental factors. For example, in arid zones, once forest cover is destroyed, the land may dry and become inhospitable to new tree growth. Other factors include overgrazing by livestock, especially animals such as goats, cows, and over-harvesting of forest resources. Together these may lead to desertification and the loss of topsoil; without soil, forests cannot grow until the long process of soil creation has been completed - if erosion allows this. In some tropical areas, forest cover removal may result in a duricrust or duripan that effectively seal off the soil to water penetration and root growth. In many areas, reforestation is impossible because people are using the land. In other areas, mechanical breaking up of duripans or duricrusts is necessary, careful and continued watering may be essential, and special protection, such as fencing, may be needed.

Countries and Regions

Afforested botanical garden in Hattori Ryokuchi Park, Japan.

Brazil

There is extensive and ongoing Amazon deforestation.

China

China has deforested most of its historically wooded areas. China reached the point where timber yields declined far below historic levels, due to over-harvesting of trees beyond sustainable yield. Although it has set official goals for reforestation, these goals are set over an 80-year time horizon and have not been significantly met by 2008. China is trying to correct these problems by projects as the Green Wall of China, which aims to replant a great deal of forests and halt the expansion of the Gobi desert. A law promulgated in 1981 requires that every school student over the age of 11 plants at least one tree per year. As a result, China has the highest afforestation rate of any country or region in the world, with 47,000 square kilometers of afforestation in 2008. However, the forest area per capita is still far lower than the international average. There has also been considerable criticism regarding the effectiveness of planting so many trees especially in regions where they never grew prior. Studies reveal that the water table of those areas is becoming deeper indicating significant water loss.

India

Afforestation in South India

India has witnessed a minor increase in the percentage of the land area under forest cover from 1950 to 2006. In 1950 around 40.48 million hectares was covered by forest. In 1980 it increased to 67.47 million hectares and in 2006 it was found to be 69 million hectares. 23% of India is covered by forest. The forests of India are grouped into 5 major categories and 16 types based on biophysical criteria. 38% of forest is categorised as subtropical dry deciduous and 30% as tropical moist deciduous plus other smaller groups. It is taken care that only local species are planted in an area. Trees bearing fruits are preferred wherever possible due to their function as a food source.

Hong Kong

Since the founding of the crown colony in the 19th century, afforestation has taken place to prevent soil erosion in the catchment areas of the reservoirs that were built. During the Japanese occupation in the Second World War, the countryside was deforested as the remaining population required fuel to survive. Most of the trees were cut down and extensive reafforestation was carried out after the war. Trees that were planted are mostly non-native species, such as: Pinus massoniana, Acacia confusa (Formosan acacia), Lophostemon confertus and the Paper Bark Tree.

Burkina Faso

Desertification is increasing along the Sahel, the strip of land between Africa's fertile tropics and the Sahara Desert. After a crippling famine in the 1970s caused by overgrazing and deforestation, a local community approach has been pioneered by Yacouba Sawadogo, a peasant farmer. By replanting trees and crops together in holes filled with compost, whole villages have been able to move back to areas considered uninhabitable.

Iran

Iran is considered a low forest cover region of the world with present cover approximating seven percent of the land area. This is a value reduced by an estimated six million hectares of virgin forest, which includes oak, almond and pistachio. Due to soil substrates, it is difficult to achieve afforestation on a large scale compared to other temperate areas endowed with more fertile and less rocky and arid soil conditions. Consequently, most of the afforestation is conducted with non-native species, leading to habitat destruction for native flora and fauna, and resulting in an accelerated loss of biodiversity.

JNF trees in the Negev Desert. Man-made dunes (here a liman) help keep in rainwater, creating an oasis.

Israel

Tree-planting is an ancient Jewish tradition, mentioned in the Talmud as being more important than greeting the Messiah. With over 240 million planted trees, Israel is one of only two countries that entered the 21st century with a net gain in the number of trees, due to massive afforestation efforts. Israeli forests are the product of a major afforestation campaign by the Jewish National Fund (JNF).

Critics argue that many JNF lands inside the West Bank were illegally confiscated from Palestinian refugees, and that the JNF furthermore should not be involved with lands in the West Bank. Shaul Ephraim Cohen has claimed that trees have been planted to restrict Bedouin herding. Susan Nathan wrote that forests were planted on the site of abandoned Arab villages after the 1948 war.

Since 2009, the JNF has provided the Palestinian Authority with 3,000 tree seedlings for a forested area being developed on the edge of the new city of Rawabi, north of Ramallah.

North Africa

In North Africa, the Sahara Forest Project coupled with the Seawater greenhouse has been proposed. Some projects have also been launched in countries as Senegal to revert desertification. As of 2010, African leaders are discussing the combining of national resources to increase effectiveness. In addition, other projects as the Keita Project in Niger have been launched in the past, and have been able to locally revert damage done by desertification.

Europe

Europe has deforested the majority of its historical forests. The European Union (EU) has paid farmers for afforestation since 1990, offering grants to turn farmland back into forest and payments for the management of forest. Between 1993 and 1997, EU afforestation policies made possible the re-forestation of over 5,000 square kilometres of land. A second program, running between 2000 and 2006, afforested more than 1000 square kilometres of land (precise statistics not yet available). A third such program began in 2007. Europe's forests are growing by 0.8 million ha a year thanks to these programmes.

In Poland, the National Program of Afforestation was introduced by the government after World War II, when area of forests shrank to 20% of country's territory. Consequently, forested areas of Poland grew year by year, and on December 31, 2006, forests covered 29% of the country. It is planned that by 2050, forests will cover 33% of Poland.

According to Food and Agriculture Organization statistics, Spain had the third fastest afforestation rate in Europe in the 1990-2005 period, after Iceland and Ireland. In those years, a total of 44,360 square kilometers were afforested, and the total forest cover rose from 13,5 to 17,9 million hectares. In 1990, forests covered 26.6% of the Spanish territory. As of 2007, that figure had risen to 36.6%. Spain today has the fifth largest forest area in the European Union.

In January 2013 the UK government set a target of 12% woodland cover in England by 2060, up from the then 10%. Government-backed initiatives such as the Woodland Carbon Code are intended to support this objective by encouraging corporations and landowners to create new woodland to offset their carbon emissions.

Alpine and Subalpine regions have undergone a lot of deforestation and then forestation in the last 300 years. Out of this has emerged much practical experience. One example is the Rotten group, which is a method to bring in stable age mixed tree communities.

Australia

In Adelaide, South Australia (a city of 1.3 million), Premier Mike Rann (2002 to 2011) launched an urban forest initiative in 2003 to plant 3 million native trees and shrubs by 2014 on 300 project sites across the metro area. The projects range from large habitat restoration projects to local biodiversity projects. Thousands of Adelaide citizens have participated in community planting days. Sites include parks, reserves, transport corridors, schools, water courses and coastline. Only trees native to the local area are planted to ensure genetic integrity. Premier Rann said the project aimed to beautify and cool the city and make it more liveable; improve air and water quality and reduce Adelaide's greenhouse gas emissions by 600,000 tonnes of Co2 a year. He said it was also about creating and conserving habitat for wildlife and preventing species loss.

United States

The United States is roughly one-third covered in forest and woodland. Nevertheless, areas in the US were subject to significant tree planting. In the 1800s people moving westward encountered the Great Plains – land with fertile soil, a growing population and a demand for timber but with few trees to supply it. So tree planting was encouraged along homesteads. Arbor Day was founded in 1872 by Julius Sterling Morton in Nebraska City, Nebraska. By the 1930s the Dust Bowl environmental disaster signified a reason for significant new tree cover. Public works programs under the New Deal saw the planting of 18,000 miles of windbreaks stretching from North Dakota to Texas to fight soil erosion.

At their summit in Copenhagen in 2009, organised by the UK based The Climate Group, leaders of subnational governments – states, regions and provinces – unanimously supported a recommendation by Premier Rann to plant one billion trees across their varied jurisdictions. The initiative was strongly supported by leaders present including Quebec Premier Jean Charest, California Governor Arnold Schwarzenegger and Scottish First Minister Alex Salmond. At a subsequent meeting in Rio de Janeiro in June 2012, The Climate Group announced that it had already received commitments by member governments to plant more than 500 million trees.

Gap Dynamics

Treefall gaps in the Amazon allow sunlight to reach the forest floor.

Gap dynamics refers to the pattern of plant growth that occurs following the creation of a forest gap, a local area of natural disturbance that results in an opening in the canopy of a forest. Gap dynamics are a typical characteristic of both temperate and tropical forests and have a wide variety of causes and effects on forest life.

Gaps are the result of natural disturbances in forests, ranging from a large branch breaking off and dropping from a tree, to a tree dying then falling over, bringing its roots to the surface of the ground, to landslides bringing down large groups of trees. Because of the range of causes, gaps, therefore, have a wide range of sizes, including small and large gaps. Regardless of size, gaps allow an increase in light as well as changes in moisture and wind levels, leading to differences in microclimate conditions compared to those from below the closed canopy, which are generally cooler and more shaded.

For gap dynamics to occur in naturally disturbed areas, either primary or secondary succession must occur. Ecological secondary succession is much more common and pertains to the process of vegetation replacement after a natural disturbance. Secondary succession results in second-growth or secondary forest, which currently covers more of the tropics than old-growth forest.

Since gaps let in more light and create diverse microclimates, they provide the ideal location and conditions for rapid plant reproduction and growth. In fact, most plant species in the tropics are dependent, at least in part, on gaps to complete their life cycles.

Disturbances

Gap dynamics are the result of disturbances within an ecosystem. There are both large scale and small scale disturbances, and both are influenced by duration and frequency. These all affect the resulting impact and regeneration patterns of the ecosystem.

Broken trees create gaps in the central Amazon.

The most common type of disturbance within a tropical ecosystem is fire. Since most nutrients in a tropical ecosystem are contained in the biomass of plants, fire is an important component of recycling these nutrients and therefore regenerating an ecosystem.

An example of a small scale disturbance is a tree falling. This can cause soil movement, which re-distributes any nutrients or organisms that were attached to the tree. The tree falling also opens up the canopy for light entrance, which can support the growth of other trees and plants.

After a disturbance, there are several ways in which regeneration can occur. One way, termed the advance regeneration pathway, is when the primary understory already contains seedlings and saplings. This method is most common in the Neotropics when faced when small scale distur-bances. The next pathway is from tree remains, or any growth from bases or roots, and is common in small disturbance gaps. The third route is referred to as the soil seed bank, and is the result of germination of seeds already found in the soil. The final regeneration pathway is the arrival of new seeds via animal dispersal or wind movement. The most critical components of the regeneration are seed distribution, germination, and survival.

Forest Gaps and Forest Regeneration

Until recently, forest regeneration practices in North America have largely followed an agricultural model, with research concentrated on techniques for establishing and promoting early growth of planted stock after clearcutting, followed by studies of growth and yield emphasizing single-species growth uninfluenced by overstorey canopy. Coates (2000) questioned this approach and proposed a shift to a more ecologically and socially based approach able to accommodate greater diversity in managed stands. Predictive models of forest regeneration and growth that take account of variable levels of canopy retention will be needed as the complexity of managed forest stands increases.

Tree regeneration occurring inside canopy gaps after disturbance has been studied widely. Studies of gap dynamics have contributed much to an understanding of the role of small-scale disturbance in forest ecosystems, but they have been little used by foresters to predict tree responses following partial cutting.

In high-latitude northern forests, position inside a gap can have a pronounced effect on resource levels (e.g., light availability) and microclimate conditions (e.g., soil temperature), especially along the north–south axis. Such variation must inevitably affect the amount and growth of regeneration; but relying solely on natural regeneration to separate the effects of gap size and position is problematic (Coates 2000). Among the many factors affecting seedling establishment following canopy disturbance are parent tree proximity and abundance, seedbed substrate, presence of seed consumers and dispersers, and climatic and microclimatic variability. Planted trees can be used to avoid many of the stochastic events surrounding natural seedling establishment.

Gradients of canopy influence can be created by partial cutting, and tree growth responses within gaps of various sizes and configurations, as well as within the adjacent forest matrix can form a basis for tree species selection. Hybrid spruce (the complex of white spruce, Sitka spruce, and occasionally Engelmann spruce) was one of several coniferous species used in a study in the Moist Cold subzone of the Interior Cedar–Hemlock zone in northwestern British Columbia. A total of 109 gaps were selected from a population of openings created by logging within each light and heavy partial cutting treatment in stands averaging 30 m in canopy height; 76 gaps were less than 1000 m², 33 were between 1000 m² and 5000 m². Canopy gap size was calculated as the area of an ellipse, the major axis of which was the longest line that could be run from canopy edge to canopy edge inside the gap, and the minor axis was the longest line that could be run from canopy edge perpendicular to the long line. Seedlings were planted in gaps and in the undisturbed and clearcut treatment units. There were strong and consistent trends in growth response among the seedlings as gap size increased. In all species, growth increased rapidly from small single-tree gaps to about 1000 m², but thereafter, there was little change up to 5000 m². Tree size and current growth rates for all species were highest in full open conditions. In large and medium gaps (300–1000 m²), the largest trees of all species occurred in the middle gap position, with little difference between the sunny north and shady south positions, lodgepole pine excepted. The light advantage expected off the north end of higher-latitude gaps was not a benefit for tree growth, suggesting that below-ground effects of canopy edge trees have an important influence of seedling growth in these forests.

In a study near Chapleau, Ontario (Groot 1992, Groot et al. 1997), openings were created in 40-year-old aspen and monitored to determine their influence on outplanted white spruce seedling development. Circular openings 9 m and 18 m in diameter, 9 m and 18 m wide east–west strips, and a 100 m × 150 m clearcut were planted and spot-seeded. The variation in solar radiation, air temperature, and soil temperature among the strips and plots was almost as great as the variation between the clearcut and intact forest. Solar radiation during the first growing season varied from 18% of the above-canopy values within the uncut stand to 68% values at the center of the 18 m strip. Near the edges of the strips, solar radiation was about 40% of the above-canopy along the south and 70% to 80% along the north. Stomatal conductance in white spruce seedlings declined generally from more sheltered to more exposed environments, correlating best with increased vapor pressure deficit (VPD). Without vegetation control, position in openings had little effect on the growth of planted white spruce; regrowth of lesser vegetation isolated seedlings from the microclimatic effects of overstorey treatment. Seedling diameters were independent of environment, while height growth was only slightly greater in environments having more light. With vegetation control, white spruce diameter and height were greatest in the center of the strips, even though there was less light there than along the north edge of the strips. Moisture stress may have accounted for that result.

Primary Succession

Succession is the slow rebuilding of forest gaps from natural or human disturbances. When major geological changes such as volcano eruptions or landslides occur, the current vegetation and soil may erode away leaving only rock. Primary succession occurs when pioneer species such as lichens colonize rock. As the lichens and mosses decompose, a soil substrate forms called peat. The peat, over time, will create a terrestrial ecosystem. From there on herbaceous, non-woody plants will develop and trees will follow. Major holes or gaps in the forest ecosystem will take hundreds of years to regenerate from a rock base.

Secondary Succession

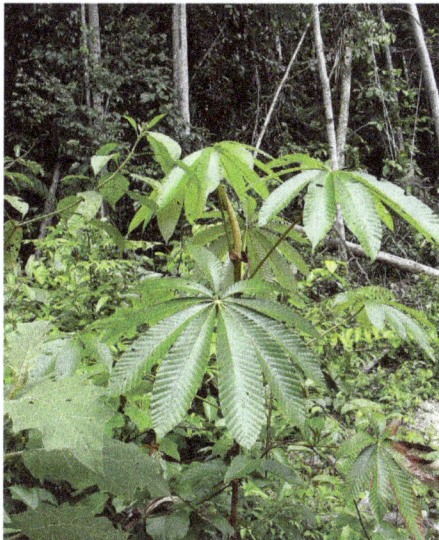

Cecropia trees are a common pioneer species found in gaps.

Secondary succession occurs where a disturbance has taken place but soil remains and is able to support plant growth. It does not take nearly as long for plant regeneration to occur because of the soil substrate already present. Secondary succession is much more common than primary succession in the tropics.

Ecological secondary succession occurs in four distinct phases: First, rapid colonization of cleared land by species such as herbs, shrubs, and climbers as well as seedlings from pioneer tree species occurs and this can last up to three years. After that, short lived but fast growing shade intolerant species form a canopy over 10 to 30 years. Non-pioneer heliophilic (sun-loving) tree species then add to the biomass and species richness as well as shade tolerant species and this can last 75 to 150 years. Finally, shade-tolerant species regain full canopy stature indefinitely until another major disturbance occurs.

Secondary succession in the tropics begins with pioneer species, which are rapidly growing and include vines and shrubs. Once these species are established, large heliophilic species will develop such as heliconias. Cecropias are also a major pioneering tree in the tropics and they are adapted to grow well where forest gaps are giving way to sunlight. Shade-tolerant species that have remained low in the forest develop and become much taller. These successional phases do not have definite order or structure and because of the very high biodiversity in the tropics, there is a lot of competition for resources such as soil nutrients and sunlight.

Examples of Tree Dynamics

Due to the fact that horizontal and vertical heterogeneity of a forest is significantly increased by gaps, gaps become an obvious consideration in explaining high biodiversity. It has been proven that gaps create suitable conditions for rapid growth and reproduction. For example, non-shade tolerant plant species and many shade-tolerant plant species respond to gaps with an increase in growth, and at least a few species are dependent on gaps to succeed in their respected environments (Brokaw 1985; Hubbell and Foster 1986b; Murray 1988; Clark and Clark 1992). Gaps create diverse microclimates, affecting light, moisture, and wind conditions (Brokaw 1985). For example, exposure to edge effects increases a microclimate's light and wind intensity and decreases its moisture. A study conducted on Barro Colorado Island in Panama showed that gaps had greater seedling establishment and higher sapling densities than control areas.

Species richness was higher in gaps than in control areas, and there was more diversity in species composition among gaps. However, this study also found that there was a low recruitment rate per gap, which explains why gaps differed in species composition. With 2% to 3% for pioneer species and 3% to 6% for shade-tolerant and intermediate species. Suggesting that most species could not take advantage of gaps because they couldn't get to them through seed dispersal. With that said, the Janzen-Connell effect plays a major role in the tree species' relationship with gaps. The Janzen-Connell density dependent mortality model states that most trees die as seed or seedlings. In addition, host-specific predators or pathogens are predicted to be greatest where density is greatest, which is underneath parent tree. This corroborates with the major causes of gaps, which are the falling of trees due to mortality caused by termites or epiphyte growth. The Janzen-Connell model also states that balance between dispersal distance and mortality should cause highest recruitment to be at a certain distance away from the parent. Therefore, if these gaps are being created by the parents, the seedlings recruit away from the gap, resulting in increasing survival rates as the distance from the parent increases. This explains the low recruitment rate per gap found in the experiment conducted in Barro Colorado Island.

In corroboration, a study conducted in La Selva in Costa Rica calculated the crown illumination index for nine tree species ranging from gap specialists to emergent canopy species. Crown illumination values ranged from 1, which indicated low light, and 6, which indicated that the tree crown was completely exposed . After using a mathematical model to calculate the changes in tree diameter and changes in crown illumination with age. This model helped estimate life expectancy, time of passage to various sizes, and age patterns of mortality. The results showed what most gap dynamics studies show, pioneer species thrived in high light environments and non-pioneer species showed high mortality when young but the rate of mortality decreased as they aged. However, once trees were very large survivorship then decreased.

Disturbance (ecology)

In biology, a disturbance is a temporary change in environmental conditions that causes a pronounced change in an ecosystem. Disturbances often act quickly and with great effect, to alter the physical structure or arrangement of biotic and abiotic elements. Disturbance can also occur over a long period of time and can impact the diversity within an ecosystem. Major ecological disturbances may include fires, flooding, windstorms, insect outbreaks and trampling. Earthquakes, various types of volcanic eruptions, tsunami, firestorms, impact events, climate change, and the devastating effects of human impact on the environment (anthropogenic disturbances) such as clearcutting, forest clearing and the introduction of invasive species can be considered major disturbances. Disturbance forces can have profound immediate effects on ecosystems and can, accordingly, greatly alter the natural community. Because of these and the impacts on populations, disturbance determines the future shifts in dominance, various species successively becoming dominant as their life history characteristics, and associated life-forms, are exhibited over time.

The disturbance of a fire can clearly be seen by comparing the unburnt (left) and burnt (right) sides of the mountain range in South Africa. The veld ecosystem relies on periodic fire disturbances like these to rejuvenate its self.

Criteria

Conditions under which natural disturbances occur are influenced mainly by climate, weather, and location. Fire disturbances will only occur in areas where there is low precipitation, some form of ignition (typically lightning), and enough flammable biomass to allow fire to spread. Conditions often occur as part of a cycle and disturbances may be periodic. Other disturbances, such as those caused by humans, invasive species or impact events, can occur anywhere and are not necessarily cyclic. Extinction vortices may result in multiple disturbances or a greater frequency of a single disturbance. Immediately after a disturbance there is a pulse of recruitment or regrowth under conditions of little competition for space or other resources. After the initial pulse, recruitment slows since once an individual plant is established it is very difficult to displace. Due to the varying forms of disturbance this directly impacts the organisms which will exploit the disturbance and create diversity within an ecosystem.

Cyclic disturbance

Often, when disturbances occur naturally, they provide conditions that favor the success of different species over pre-disturbance organisms. This can be attributed to physical changes in the biotic and abiotic conditions of an ecosystem. Because of this, a disturbance force can change an ecosystem for significantly longer than the period over which the immediate effects persist. With the passage of time following a disturbance, shifts in dominance may occur with ephemeral herbaceous life-forms progressively becoming over topped by taller perennials herbs, shrubs and trees. However, in the absence of further disturbance forces, many ecosystems trend back toward pre-disturbance conditions. Long lived species and those that can regencrate in the presence of their own adults finally become dominant. Such alteration, accompanied by changes in the abundance of different species over time, is called ecological succession. Succession often leads to conditions that will once again predispose an ecosystem to disturbance.

Damages of storm Kyrill in Wittgenstein, Germany.

Pine forests in the western North America provide a good example of such a cycle involving insect outbreaks. The mountain pine beetle (*Dendroctonus ponderosae*) play an important role in limiting pine trees like lodgepole pine in forests of western North America. In 2004 the beetles affected more than 90,000 square kilometres. The beetles exist in endemic and epidemic phases. During epidemic phases swarms of beetles kill large numbers of old pines. This mortality creates openings in the forest for new vegetation. Spruce, fir, and younger pines, which are unaffected by the beetles, thrive in canopy openings. Eventually pines grow into the canopy and replace those lost. Younger pines are often able to ward off beetle attacks but, as they grow older, pines become less vigorous and more susceptible to infestation. This cycle of death and re-growth creates a temporal mosaic of pines in the forest. Similar cycles occur in association with other disturbances such as fire and windstorms.

When multiple disturbance events affect the same location in quick succession, this often results in a "compound disturbance," an event which, due to the combination of forces, creates a new situation which is more than the sum of its parts. For example, windstorms followed by fire can create fire temperatures and durations that are not expected in even severe wildfires, and may have surprising effects on post-fire succession. Environmental stresses can be described as pressure on the environment, with compounding variables such as extreme temperature or precipitation changes—which all play a role in the diversity and succession of an ecosystem. With environmental moderation, diversity increases because of the intermediate- disturbance effect, decreases because of the competitive-exclusion effect, increases because of the prevention of competitive exclusion by moderate predation, and decreases because of the local extinction of prey by severe predation. A reduction in recruitment density reduces the importance of competition for a given level of environmental stress.

Species Adapted to Disturbance

A disturbance may change a forest significantly. Afterwards, the forest floor is often littered with dead material. This decaying matter and abundant sunlight promote an abundance of new growth. In the case of forest fires a portion of the nutrients previously held in plant biomass is returned quickly to the soil as biomass burns. Many plants and animals benefit from disturbance conditions. Some species are particularly suited for exploiting recently disturbed sites. Vegetation with the potential for rapid growth can quickly take advantage of the lack of competition. In the northeastern United States, shade-intolerant trees like pin cherry and aspen quickly fill in forest gaps created by fire or windstorm (or human disturbance). Silver maple and eastern sycamore are similarly well adapted to floodplains. They are highly tolerant of standing water and will frequently dominate floodplains where other species are periodically wiped out.

Forest fire burns on the island of Zakynthos in Greece on July 25th, 2007.

When a tree is blown over, gaps typically are filled with small herbaceous seedlings but, this is not always the case; shoots from the fallen tree can develop and take over the gap. The sprouting ability can have major impacts on the plant population, plant populations that typically would have exploited the tree fall gap get over run and can not complete against the shoots of the fallen tree. Species adaptation to disturbances is species specific but how each organism adapts effects all the species around them.

Another species well adapted to a particular disturbance is the Jack Pine in boreal forests exposed to crown fires. They, as well as some other pine species, have specialized serotinous cones that only open and disperse seeds with sufficient heat generated by fire. As a result, this species often dominates in areas where competition has been reduced by fire.

Species that are well adapted for exploiting disturbance sites are referred to as pioneers or early successional species. These shade-intolerant species are able to photosynthesize at high rates and as a result grow quickly. Their fast growth is usually balanced by short life spans. Furthermore, although these species often dominate immediately following a disturbance, they are unable to compete with shade-tolerant species later on and replaced by these species through succession. However these shifts may not reflect the progressive entry to the community of the taller long-lived forms, but instead, the gradual emergence and dominance of species that may have been present, but inconspicuous directly after the disturbance.

While plants must deal directly with disturbances, many animals are not as immediately affected by them. Most can successfully evade fires, and many thrive afterwards on abundant new growth on the forest floor. New conditions support a wider variety of plants, often rich in nutrients compared to pre-disturbance vegetation. The plants in turn support a variety of wildlife, temporarily increasing biological diversity in the forest.

Importance

Biological diversity is dependent on natural disturbance. The success of a wide range of species from all taxonomic groups is closely tied to natural disturbance events such as fire, flooding, and windstorm. As an example, many shade-intolerant plant species rely on disturbances for success-ful establishment and to limit competition. Without this perpetual thinning, diversity of forest flora can decline, affecting animals dependent on those plants as well.

A good example of this role of disturbance is in ponderosa pine (*Pinus ponderosa*) forests in the western United States, where surface fires frequently thin existing vegetation allowing for new growth. If fire is suppressed, douglas fir (*Pesudotsuga menziesii*), a shade tolerant species, even-tually replaces the pines. Douglas firs, having dense crowns, severely limit the amount of sunlight reaching the forest floor. Without sufficient light new growth is severely limited. As the diversity of surface plants decreases, animal species that rely on them diminish as well. Fire, in this case, is important not only to the species directly affected but also to many other organisms whose survival depends on those key plants.

Diversity is low in harsh environments because of the intolerance of all but opportunistic and highly resistant species to such conditions. The interplay between disturbance and these biological processes seems to account for a major portion of the organization and spatial patterning of nat-ural communities. Disturbance variability and species diversity are heavily linked, and as a result require adaptations that help increase plant fitness necessary for survival.

Reforestation

Reforestation is the natural or intentional restocking of existing forests and woodlands that have been depleted, usually through deforestation. Reforestation can be used to rectify or im-prove the quality of human life by soaking up pollution and dust from the air, rebuild natural habitats and ecosystems, mitigate global warming since forests facilitate biosequestration of atmospheric carbon dioxide, and harvest for resources, particularly timber, but also non-tim-ber forest products.

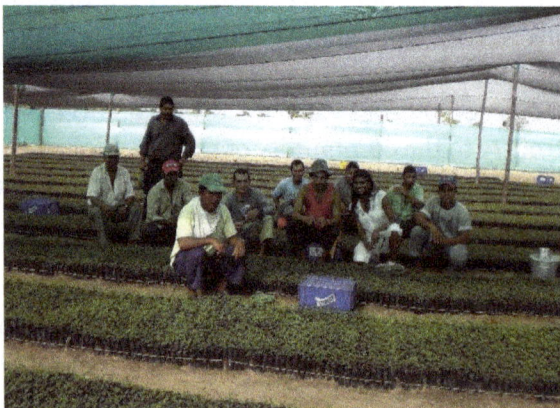

Tropical tree nursery at Planeta Verde Reforestación S.A.'s plantation in Vichada Department, Colombia

A 21-year-old plantation of red pine in Southern Ontario

The term reforestation is similar to afforestation, the process of restoring and recreating areas of woodlands or forests that may have existed long ago but were deforested or otherwise removed at some point in the past. Sometimes the term *re-afforestation* is used to distinguish between the original forest cover and the later re-growth of forest to an area. Forestation is the establishment of forest growth on areas that either had forest or lacked it.

Special tools, e.g. tree planting bars, are used to make planting of trees easier and faster.

Management

A debated issue in managed reforestation is whether or not the succeeding forest will have the same biodiversity as the original forest. If the forest is replaced with only one species of tree and all other vegetation is prevented from growing back, a monoculture forest similar to agricultural crops would be the result. However, most reforestation involves the planting of different seedlots of seedlings taken from the area, often of multiple species. Another important factor is the natural regeneration of a wide variety of plant and animal species that can occur on a clear cut. In some areas the suppression of forest fires for hundreds of years has resulted in large single aged and single species forest stands. The logging of small clear cuts and/or prescribed burning, actually increases the biodiversity in these areas by creating a greater variety of tree stand ages and species.

For Harvesting

Reforestation need not be only used for recovery of accidentally destroyed forests. In some countries, such as Finland, many of the forests are *managed* by the wood products and pulp and paper industry. In such an arrangement, like other crops, trees are planted to replace those that have been cut. In such circumstances, the industry can cut the trees in a way to allow easier reforestation. The wood products industry systematically replaces many of the trees it cuts, employing large numbers of summer workers for tree planting work. For example, in 2010, Weyerhaeuser reported planting 50 million seedlings. However replanting an old-growth forest with a plantation is not replacing the old with the same characteristics in the new.

In just 20 years, a teak plantation in Costa Rica can produce up to about 400 m³ of wood per hectare. As the natural teak forests of Asia become more scarce or difficult to obtain, the prices commanded by plantation-grown teak grows higher every year. Other species such as mahogany grow more slowly than teak in Tropical America but are also extremely valuable. Faster growers include pine, eucalyptus, and *Gmelina*.

Reforestation, if several indigenous species are used, can provide other benefits in addition to financial returns, including restoration of the soil, rejuvenation of local flora and fauna, and the capturing and sequestering of 38 tons of carbon dioxide per hectare per year.

The reestablishment of forests is not just simple tree planting. Forests are made up of a community of species and they build dead organic matter into soils over time. A major tree-planting program could enhance the local climate and reduce the demands of burning large amounts of fossil fuels for cooling in the summer.

For Climate Change Mitigation

Forests are an important part of the global carbon cycle because trees and plants absorb carbon dioxide through photosynthesis. By removing this greenhouse gas from the air, forests function as terrestrial carbon sinks, meaning they store large amounts of carbon. At any time, forests account for as much as double the amount of carbon in the atmosphere. Even as more anthropogenic carbon is produced, forests remove around three billion tons of anthropogenic carbon every year. This amounts to about 30% of all carbon dioxide emissions from fossil fuels. Therefore, an increase in the overall forest cover around the world would tend to mitigate global warming.

There are four major strategies available to mitigate carbon emissions through forestry activities: increase the amount of forested land through a reforestation process; increase the carbon density of existing forests at a stand and landscape scale; expand the use of forest products that will sustainably replace fossil-fuel emissions; and reduce carbon emissions that are caused from deforestation and degradation.

Achieving the first strategy would require enormous and wide-ranging efforts. However, there are many organizations around the world that encourage tree-planting as a way to offset carbon emissions for the express purpose of fighting climate change. For example, in China, the Jane Goodall Institute, through their Shanghai Roots & Shoots division, launched the Million Tree Project in Kulun Qi, Inner Mongolia to plant one million trees to stop desertification and help curb climate change. China has used 24 billion metres squared of new forest plantation and natural forest regrowth to offset 21% of Chinese fossil fuel emissions in 2000. In Java, Indonesia each newlywed couple is to give whoever is sermonizing their wedding 5 seedlings to combat global warming. Each couple that wishes to have a divorce has to give 25 seedlings to whoever divorces them.

The second strategy has to do with selecting species for tree-planting. In theory, planting any kind of tree to produce more forest cover would absorb more carbon dioxide from the atmosphere. On the other hand, a genetically modified tree specimen might grow much faster than any other regular tree. Some of these trees are already being developed in the lumber and biofuel industries. These fast-growing trees would not only be planted for those industries but they can also be planted to help absorb carbon dioxide faster than slow-growing trees.

Extensive forest resources placed anywhere in the world will not always have the same impact. For example, large reforestation programs in boreal or subarctic regions have a limited impact on climate mitigation. This is because it substitutes a bright snow-dominated region that reflects the sunlight with dark forest canopies. A study from the National Center for Atmospheric Research in Boulder, Colorado, USA, found that trees in temperate latitudes have a net warming effect on the atmosphere. The heat that dark leaves release without absorbing outweighs the carbon they sequester. On the other hand, a positive example would be reforestation projects in tropical regions, which would lead to a positive biophysical change such as the formation of clouds. These clouds would then reflect the sunlight, creating a positive impact on climate mitigation.

There is an advantage to planting trees in tropical climates with wet seasons. In such a setting, trees have a quicker growth rate because they can grow year-round. Trees in tropical climates have, on average, larger, brighter, and more abundant leaves than non-tropical climates. A study of the girth of 70,000 trees across Africa has shown that tropical forests are soaking up more carbon dioxide pollution than previously realized. The research suggests almost one fifth of fossil fuel emissions are absorbed by forests across Africa, the Amazon basin and Asia. Simon Lewis, a climate expert at the University of Leeds, who led the study, said: "Tropical forest trees are absorbing about 18% of the carbon dioxide added to the atmosphere each year from burning fossil fuels, substantially buffering the rate of change."

It is also important to deal with the rate of deforestation. At this point, there are 13 billion metres squared of tropical regions that are deforested every year. There is potential for these regions to reduce rates of deforestation by 50% by 2050, which would be a huge contribution to stabilize the global climate.

On Abandoned Farmland

With increased intensive agriculture and urbanization, there is an increase in the amount of abandoned farmland. By some estimates, for every half a hectare of original old-growth forest cut down, more than 20 hectares of new secondary forests are growing, even though they do not have the same biodiversity as the original forests and original forests store 60% more carbon than these new secondary forests. According to a study in Science, promoting regrowth on abandoned farmland could offset years of carbon emissions.

Promotion Strategies

Land Rights

Transferring land rights to indigenous inhabitants is argued to efficiently conserve forests.

Transferring rights over land from public domain to its indigenous inhabitants is argued to be a cost effective strategy to conserve forests. This includes the protection of such rights entitled in existing laws, such as India's Forest Rights Act. The transferring of such rights in China, perhaps the largest land reform in modern times, has been argued to have increased forest cover. In Brazil, forested areas given tenure to indigenous groups have even lower rates of clearing than national parks.

Incentives

Some incentives for reforestation can be as simple as a financial compensation. Streck and Scholz (2006) explain how a group of scientists from various institutions have developed a compensated reduction of deforestation approach which would reward developing countries that disrupt any further act of deforestation. Countries that participate and take the option to reduce their emissions from deforestation during a committed period of time would receive financial compensation for the carbon dioxide emissions that they avoided. To raise the payments, the host country would issue government bonds or negotiate some kind of loan with a financial institution that would want to take part in the compensation promised to the other country. The funds received by the country could be invested to help find alternatives to the extensive cutdown of forests. This whole process of cutting emissions would be voluntary, but once the country has agreed to lower their emissions they would be obligated to reduce their emissions. However, if a country was not able to meet their obligation, their target would get added to their next commitment period. The authors of these proposals see this as a solely government-to-government agreement; private entities would not participate in the compensation trades.

Examples

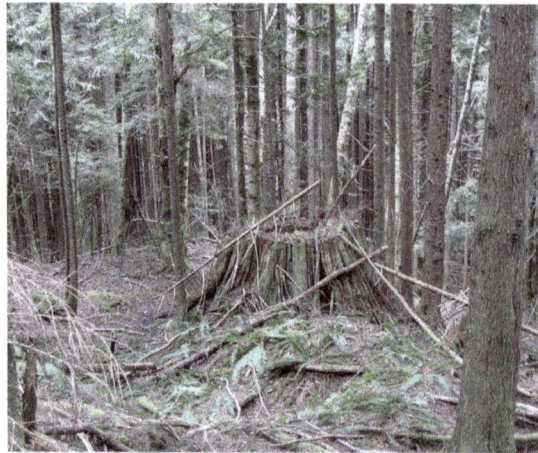

Forest regrowth in Mount Baker-Snoqualmie National Forest, Washington state, USA

Sub-Saharan Africa

Plans to plant a nine-mile width of trees on the Southern Border of the Sahara Desert. The Great Green Wall initiative is a pan-African proposal to "green" the continent from west to east in order to battle desertification. It aims at tackling poverty and the degradation of soils in the Sahel-Saharan region, focusing on a strip of land of 15 km (9 mi) wide and 7,500 km (4,750 mi) long from Dakar to Djibouti.

Canada

In Canada, overall forest cover is increasing over the last few decades.

China

In China, extensive replanting programs have existed since the 1970s. Programs have had overall success. The forest cover has increased from 12% of China's land area to 16%. However, specific programs have had limited success. The "Green Wall of China", an attempt to limit the expansion of the Gobi Desert is planned to be 2,800 miles (4,500 km) long and to be completed in 2050. China plans to plant 26 billion trees in the next decade that is two trees for every Chinese citizen per year. China requires that students older than 11 years old plant one tree a year until their high school graduation.

Germany

Reforestation is required as part of the federal forest law. 31% of Germany is forested, according to the second forest inventory of 2001–2003. The size of the forest area in Germany increased between the first and the second forest inventory due to forestation of degenerated bogs and agricultural areas.

India

Jadav Payeng had received national awards for reforestation efforts, known as the "Molai forest".

Japan

Mass environmental and human-body pollution along with relating deforestation, water pollution, smoke damage, and loss of soils caused by mining operations in Ashio, Tochigi became the first environmental social issue in Japan, efforts by Shōzō Tanaka had grown to large campaigns against copper operation. This led to the creation of 'Watarase Yusuichi Pond', to settle the pollution which is a Ramsar site today. Reforestation was conducted as a part of afforestation due to inabilities of self-recovering by the natural land itself due to serious soil pollution and loss of woods consequence in loss of soils for plants to grow, thus needing artificial efforts involving introducing of healthy soils from outside. Starting since in 1897, about 50% of once bald mountains backed to green.

Lebanon

For thousands of years Lebanon was covered by forests, one particular species of interest, *Cedrus libani* or Lebanon Cedar was exceptionally valuable timber species. Virtually every ancient culture that shared the Mediterranean Sea harvested these trees, the Ancient Egyptians, the Greeks, the Roman Empire, early Christians, the Muslims, Persians, Assyrians and Babylonians. Extensive deforestation has occurred, with only small remnants of the original forests surviving. Deforestation has been particularly severe in Lebanon and on Cyprus. The Lebanon Reforestation Initiative aims to restore Lebanon's native forests. Projects financed locally and by international charity are performing extensive reforestation of cedar being carried out in the Mediterranean region, particularly in Lebanon and Turkey, where over 50 million young cedars are being planted annually.

Philippines

The Philippines increased its forest area to 240,000 hectares (593,053 acre) per year from 2010 to 2015. In 2011-2015 period, reforestation through the government's National Greening Program covered 1,258,692 hectares of land, which is 105 percent more than the 1,200,000 hectares planting target for the same period. An estimated eight million hectares of land nationwide need reforestation and rehabilitation, according to the Forest Management Bureau.

United States

It is the stated goal of the US Forest Service to manage forest resources sustainably. This includes reforestation after timber harvest, among other programs.

Organizations

Trees for the Future has assisted more than 170,000 families, in 6,800 villages of Asia, Africa and the Americas, to plant over 35 million trees.

Wangari Maathai, 2004 Nobel Peace Prize recipient, founded the Green Belt Movement which planted over 47 million trees to restore the Kenyan environment.

Shanghai Roots & Shoots, a division of the Jane Goodall Institute, launched The Million Tree Project in Kulun Qi, Inner Mongolia to plant one million trees to stop desertification and alleviate global warming.

Criticisms

Competition with Human Interests

A farmland near the Bennets Woodland, near Stichens Green and Streatley, Bedfordshire, the United Kingdom, 2005.

Reforestation competes with other land uses such as food production, livestock grazing and living space for further economic growth. However, in the case of food production, data from the Food and Agriculture Organization shows that between 1961 and 2012, the amount of arable land needed to produce the same amount of food declined by 68 percent worldwide.

Reforestation often has the tendency to create large fuel loads, resulting in significantly hotter combustion than fires involving low brush or grasses. It can divert large amounts of water from other activities. It sometimes results in extensive canopy creation that prevents growth of diverse vegetation in the shadowed areas and generating soil conditions that hamper other types of vegetation. Trees used in some reforesting efforts (e.g., *Eucalyptus globulus*) tend to extract large amounts of moisture from the soil, preventing the growth of other plants. An increase in the number of humans globally means that more food would have to be produced from farms to feed them. However, the growth rate is currently declining.

Natural Risks

There is also the risk that through a forest fire or insect outbreak much of the stored carbon in a reforested area could make its way back to the atmosphere. Reduced harvesting rates and fire suppression have caused an increase in the forest biomass in the western part of the Continental United States, over the past century. This causes an increase of about a factor of four in the frequency of fires due to longer and hotter dry seasons.

Urban Forestry

Tree pruning in Durham, North Carolina

James Kinder, an ISA Certified Municipal Arborist examining a Japanese Hemlock at Hoyt Arboretum

Urban forestry is the care and management of single trees and tree populations in urban settings for the purpose of improving the urban environment. Urban forestry advocates the role of trees as a critical part of the urban infrastructure. Urban foresters plant and maintain trees, support appropriate tree and forest preservation, conduct research and promote the many benefits trees provide. Urban forestry is practiced by municipal and commercial arborists, municipal and utility foresters, environmental policymakers, city planners, consultants, educators, researchers and community activists.

Functions and Values

Function, the dynamic operation of the forest, includes biochemical cycles, gas exchange, primary productivity, competition, succession, and regeneration. In urban environments, forest functions are frequently related to the human environment. Trees are usually selected, planted, trimmed, and nurtured by people, often with specific intentions, as when a tree is planted in a front yard to shade the driveway and frame the residence. The functional benefits provided by this tree depend on structural attributes, such as species and location, as well as management activities that influence its growth, crown dimensions, and health.

Urban forest functions are thus often oriented toward human outcomes, such as shade, beauty, and privacy. As prominent "things," arranged in distinctive formations, trees command a symbolic and material presence that informs how places and landscapes are imagined. This link that humans have to trees has been theorized by Kellert and Wilson (1993) to be a genetically based emotional need to be close to trees and other greenery. According to their "Biophilia Hypothesis," millions of years of human survival and evolution depended on our ability to cope with the natural world; learning what was safe and dangerous involved the imprinting of strong positive and negative emotional reactions to various natural stimuli. Although 21st century American society is no longer as dependent on nature for day-to-day survival, Kellert and Wilson suggest that closeness to the natural world is still critical for psychological well-being. The complex symbolic and emotional ties that humans have with trees have important implications for the importance of sound urban forest management practices that impact not only quality of life on an ecological level, but on the human and cultural level. People develop emotional attachments to trees that give them special status and value. Removing "hazardous" trees can be difficult when it means severing the connection between residents and the trees they love. For many, feelings of attachment to trees in cities influences feelings for preservation of trees in forests (McPherson 1998).

Professional Tree Climber (arborist: Zack Weiler) climbing a willow tree in Port Elgin, ON. Canada

The value that people place on trees is especially evident with respect to big trees. There has always been a public fascination with large trees, especially the largest specimens of trees that reach a mature height of greater than 40 or 50 feet (i.e., Champion Trees) (Barro et al. 1997, Dwyer et al. 1991). Moreover, the ability of big street trees to create a ceiling of branches and leaves over all or part of a street impacts the scale of changing shadows cast by the trees, sunlight filtration, and other human-scale considerations that provide a changing visual environment (Zube 1973, Jones and Cloke 2002). In their qualitative study of Denmark residents' perceptions of the importance of the urban forest, Hansen-Moller and Oustrup (2004) found that the scale of urban trees was one of the main conditions of an "ideal" urban forest, through its volume, height, and ability to envelop a person, thus creating a barrier from the outside world.

Urban forests bring many environmental and economic benefits to cities. Among these are energy benefits in the form of reduced air conditioning by shading buildings, homes and roads, absorbing sunlight, reducing ultraviolet light, cooling the air, and reducing wind speed - in short improvement of the microclimate and air quality (McPherson 1994; McPherson & Rowntree 1993; Simpson & McPherson 1996; Coder 1996; Wolfe 1999; Hastie 2003; Lohr *et al.* 2004). There are also economic benefits associated with urban trees such as increased land, property, and rental value (Morales et al. 1983; Anderson & Cordell 1988; Wolf 1998; Dwyer et al. 1992; Mansfield et al. 2005; Orland et al. 1992; Hastie 2003; USDA Forest Service 2003, 2004). Well-maintained trees and landscaped business districts have been shown to encourage consumer purchases and attract increased residential, commercial and public investments (Wolf 2004, 2007). Trees located in business areas may also increase worker productivity, recruitment, retention and satisfaction (Kaplan & Kaplan 1989; Kaplan 1992; Wolf 1998). Urban forests also improve air quality, absorb rainwater, improve biodiversity and potentially allow recycling to 20% of waste which is wood-based

Many cities today are dealing with stormwater management system issues where their existing systems can no longer hold the volume of water that falls in storms. One sustainable solution to this is planting street trees with grates underneath them to hold water. Trees and their soils work to filter runoff pollution and soil contaminants by absorbing them and processing them into less harmful substances. They also collect water in their limbs and release it back into the atmosphere over time. This makes trees a solution to stormwater runoff issues and urban heating issues.

The social and even medical benefits of nature are also dramatic. Urban poverty is common to areas lacking green spaces. Visiting green areas in cities can counteract the stress of city life, renew vital energy and restore attention, and improve medical outcomes. Simply being able to see a natural view out of the window improves self-discipline in inner city girls.

Having regular access to woodland is desirable for schools, and indeed Forest kindergartens take children to visit substantial forests every day, whatever the weather. When such children go to primary school, teachers observe a significant improvement in reading, writing, mathematics, social skills and many other areas.

Various methods are available to capture the value of urban trees, each designed to analyse a specific type of green space (individual trees, parks, trees on golf courses etc.). The following are examples of studies that have used these different approaches, along with their respective constraints.

Method	Study	Location	Results	Limitations
Contingent valuation	Tyrvainen (2001)	Joensuu and Salo, Finland	More than two-thirds of the respondents were willing to pay for the use of recreation areas, with mean WTP ranging from 42 to 53 FIM/ month, depending on their location.	Estimated value of environmental amenities is based on a hypothetical market scenario
Choice modeling and survey	Salazar and Menendez (2007)	Valencia, Spain	Residents closer to a proposed park had a higher WTP for the park than those further from it.	Bias, protest answers, strategic answers
Direct estimates	Pandit and Laband (2010)	Auburn, Alabama, USA	17.5 percent tree cover on property = 14.4 percent reduction in electricity ($31/month) 50 percent dense shade = 19.3 percent reduction in electricity ($42/month).	Mitigation effects of climate excluded, Aesthetic values excluded
Numerical modeling	McPherson et al. (2005)	USA	Every dollar invested in urban tree management returned annual benefits ranging from $1.37 to $3.09.	Aesthetic values excluded

Practice

Urban forestry is a practical discipline, which includes tree planting, care, and protection, and the overall management of trees as a collective resource. The urban environment can present many arboricultural challenges such as limited root and canopy space, poor soil quality, deficiency or excess of water and light, heat, pollution, mechanical and chemical damage to trees, and mitigation of tree-related hazards. Among those hazards are mostly non-immediate risks like the probability that individual trees will not withstand strong winds (as during a thunderstorm) and damage parking cars or injure passing pedestrians. Although quite striking in an urban environment, large trees in particular present a continuing dilemma for the field of urban forestry due to the stresses that urban trees undergo from automobile exhaust, constraining hardscape and building foundations, and physical damage (Pickett et al. 2008). Urban forestry also challenges the arborists that tend the trees. The lack of space requires greater use of rigging skills and traffic and pedestrian control. The many constraints that the typical urban environment places on trees limits the average lifespan of a city tree to only 32 years – 13 years if planted in a downtown area – which is far short of the 150-year average life span of trees in rural settings (Herwitz 2001).

Management challenges for urban forestry include maintaining a tree and planting site inventory, quantifying and maximizing the benefits of trees, minimizing costs, obtaining and maintaining public support and funding, and establishing laws and policies for trees on public and on private land. Urban forestry presents many social issues that require addressing to allow urban forestry to be seen by the many as an advantage rather than a curse on their environment. Social issues include under funding which leads to inadequate maintenance of urban trees. In the UK the National Urban Forestry Unit produced a series of case studies around best practice in urban forestry which is archived here.

By Country

United States

Tree warden laws in the New England states are important examples of some of the earliest and most far-sighted state urban forestry and forest conservation legislation. In 1896, the Massachusetts legislature passed the first tree warden law, and the other five New England states soon followed suit: Connecticut, Rhode Island, and New Hampshire in 1901, Vermont in 1904, and Maine in 1919. (Kinney 1972, Favretti 1982, Campanella 2003).

As villages and towns grew in population and wealth, ornamentation of public, or common, spaces with shade trees also increased. However, the ornamentation of public areas did not evolve into a social movement until the late 18th century, when private individuals seriously promoted and sponsored public beautification with shade and ornamental trees (Favretti 1982, Lawrence 1995). Almost a century later, around 1850, institutions and organization were founded to promote ornamentation through private means (Egleston 1878, Favretti 1982). In the 1890s, New England's "Nail" laws enabled towns to take definitive steps to distinguish which shade trees were public. Chapter 196 of the 1890 Massachusetts Acts and Resolves stated that a public shade tree was to be designated by driving a nail or spike, with the letter M plainly impressed on its head, into the relevant trunk. Connecticut passed a similar law in 1893, except its certified nails and spikes bore the letter C. (Northrup 1887).

The rapid urbanization of American cities in the late 19th century was a concern to many as encouraging intellectual separation of humanity and nature (Rees 1997). By the end of the 19th century, social reformers were just beginning to understand the relationship between developing parks in urban areas and "[engendering] a better society" (Young 1995:536). At this time, parks and trees were not necessarily seen as a way to allow urban dwellers to experience nature, but more of a means of providing mechanisms of acculturation and control for newly arrived immigrants and their children (e.g., areas to encourage "structured play" and thus serve as a deterrent for youth crime) (Pincetl and Gearin 2005). Other prominent public intellectuals were interested in exploring the synergy between ecological and social systems, including American landscape architect Fredrick Law Olmsted, designer of 17 major U.S. urban parks and a visionary in seeing the value of including green space and trees as a fundamental part of metropolitan infrastructure (Young 2009). To Olmsted, unity between nature and urban dwellers was not only physical, but also spiritual: "Gradually and silently the charm comes over us; the beauty has entered our souls; we know not exactly when or how, but going away we remember it with a tender, subdued, filial-like joy" (Beveridge and Schuyler 1983 cited in Young 2009:320). The conscious inclusion of trees in urban designs for American cities such as Chicago, San Francisco, and Minneapolis was also inspired by Paris's urban forest and its broad, tree-lined boulevards as well as by the English romantic landscape movement (Zube 1973). The belief in green cover by early park proponents as a promoter of social cohesion has been corroborated by more recent research that links trees to the presence of stronger ties among neighbors, more adult supervision of children in outdoor areas, more use of the neighborhood common areas, and fewer property and violent crime (Kuo et al. 1998, Kuo and Sullivan 2001, Kuo 2003).

Many municipalities throughout the United States employ community-level tree ordinances to empower planning officials to regulate the planting, maintenance, and preservation of trees. The development of tree ordinances emerged largely as a response to the Dutch Elm Disease that plagued cities from the 1930s to 1960s, and grew in response to urban development, loss of urban tree canopy, and rising public concern for the environment (Wolf 2003). The 1980s saw the beginning of the second generation of ordinances with higher standards and specific foci, as communities sought to create more environmentally pleasing harmony between new development and existing infrastructure. These new ordinances, legislated by local governments, may include specific provisions such as the diameter of tree and percentage of trees to be protected during construction activities (Xiao 1995). The implementation of these tree ordinances is greatly aided by a significant effort by community tree advocates to conduct public outreach and education aimed at increasing environmental concern for urban trees, such as through National Arbor Day celebrations and the USDA Urban and Community Forestry Program (Dwyer et al. 2000, Hunter and Rinner 2004, Norton and Hannon 1997, Wall et al. 2006). Much of the work on the ground is performed by non-profits funded by private donations and government grants.

Policy on urban forestry is less contentious and partisan than many other forestry issues, such as resource extraction in national forests. However, the uneven distribution of healthy urban forests across the landscape has become a growing concern in the past 20 years. This is because the urban forest has become an increasingly important component of bioregional ecological health with the expanding ecological footprint of urban areas. Based on American Forests' Urban Ecosystem Analyses conducted over the past six years in ten cities, an estimated 634,407,719 trees have been lost from metropolitan areas across the U.S. as the result of urban and suburban development (American Forests 2011). This is often due to the failure of municipalities to integrate trees and other elements of the green infrastructure into their day-to-day planning and decision-making processes (American Forests 2002). The inconsistent quality of urban forestry programs on the local level ultimately impacts the regional context in which contiguous urban forests reside, and is greatly exacerbated by suburban sprawl as well as other social and ecological effects (Webb et al. 2008). The recognition of this hierarchical linkage among healthy urban forests and the effectiveness of broader ecosystem protection goals (e.g., maintaining biodiversity and wildlife corridors), highlights the need for scientists and policymakers to gain a better understanding of the socio-spatial dynamics that are associated with tree canopy health at different scales (Wu 2008).

United Kingdom

In the UK urban forestry was pioneered around the turn of the 19th century by the Midland reafforesting association, whose focus was in the Black Country. In the mid 1990s the National Urban Forestry Unit (NUFU) grew out of a Black Country Urban Forestry Unit and promoted urban forestry across the UK, notably including the establishment of the Black Country Urban Forest. As urban forestry become more mainstream in the 21st century, NUFU was wound up, and its advocacy role now carried on by organisations such as The Wildlife Trusts and the Woodland Trust.

Toronto

Toronto is a diverse city with a mosaic urban forest – a patchwork of unique situations made up of trees growing in the many residential yards, lining the public streets, and beautifying public parks. Unlike the trees that grow in a wild setting, urban trees are faced with harsh conditions that can be detrimental to their health and growing potential. Soil compaction, air pollution, habitat fragmentation and competition from invasive species are some of the hardships city trees endure. Some neighbourhoods have a geriatric tree population; many mature trees that will reach the end of their lifespan very soon, with few young trees to replace them.

Some neighbourhoods suffer a serious lack of species diversity, with mainly ornamental, non-native or invasive tree species such as Bradford pear, Japanese tree lilac and Norway maple. Still other neighbourhoods, most often newly constructed subdivisions, lack tree cover completely.

Simply planting more trees cannot solve the problems faced by the urban forest. Through creative and innovative approaches the public, partnered with private enterprises can maximize the potential benefits of trees planted, and minimize the stresses they will have to overcome.

Although most people express a concern for urban trees and consider them very important, many lack the basic knowledge and skills needed to address and prevent the issues listed above. Collective action, or inaction, will make or break the future of the urban forest. Through fostering a sense of ownership amongst Toronto residents for this commonly owned resource, residents will enjoy better air quality and reduce their demand for energy.

Constraints

Resolving limitations will require coordinated efforts among cities, regions, and countries (Meza, 1992; Nilsson, 2000; Valencia, 2000).

- Loss of green space is continuous as cities expand; available growing space is limited in city centres. This problem is compounded by pressure to convert green space, parks, etc. into building sites (Glickman, 1999).

- Inadequate space is allowed for the root system.

- Poor soil is used when planting specimens.

- Incorrect and neglected staking leads to bark damage.

- Larger, more mature trees are often used to provide scale and a sense of establishment to a scheme. These trees grow more slowly and do not thrive in alien soils whilst smaller specimens can adapt more readily to existing conditions.

- Lack of information on the tolerances of urban tree cultivars to environmental constraints.

- Poor tree selection which leads to problems in the future

- Poor nursery stock and failure of post-care

- Limited genetic diversity

- Too few communities have working tree inventories and very few have urban forest management plans.

- Lack of public awareness about the benefits of healthy urban forests.

- Poor tree care practices by citizens and untrained arborists.

Organizations

- Alliance for Community Trees
- American Forests
- Casey Trees
- Friends of the Urban Forest
- Greening of Detroit
- Hantz Woodlands
- International Society of Arboriculture
- National Urban Forestry Unit
- OpenTreeMap
- Our City Forest
- Society of Municipal Arborists

- Society of American Foresters
- LEAFTennessee Urban Forestry Council
- The Tree Council, UK
- Tree City USA Program
- TREE Fund
- TreeLink
- Trees Are Good
- Canopy.org
- Trees Atlanta
- Arborist Video Blog
- TreesCharlotte

Biosequestration

Biosequestration is the capture and storage of the atmospheric greenhouse gas carbon dioxide by biological processes.

This may be by increased photosynthesis (through practices such as reforestation / preventing deforestation and genetic engineering); by enhanced soil carbon trapping in agriculture; or by the use of algal bio sequestration to absorb the carbon dioxide emissions from coal, petroleum (oil) or natural gas-fired electricity generation.

Flowering Corymbia ficifolia, Austins Ferry, Tasmania, Australia

Biosequestration as a natural process has occurred in the past, and was responsible for the formation of the extensive coal and oil deposits which are now being burned. It is a key policy concept in the climate change mitigation debate. It does not generally refer to the sequestering of carbon dioxide in oceans or rock formations, depleted oil or gas reservoirs, deep saline aquifers, or deep coal seams or through the use of industrial chemical carbon dioxide scrubbing.

The importance of Plants in Storing Atmospheric Carbon Dioxide

Kew Gardens Waterlily House. David Iliff, 2008

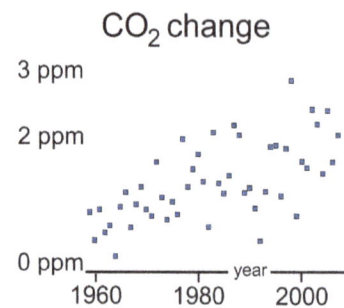

Recent year-to-year increase of atmospheric CO_2

After water vapour (concentrations of which humans have limited capacity to influence) carbon dioxide is the most abundant and stable greenhouse gas in the atmosphere (methane rapidly reacts to form water vapour and carbon dioxide). Atmospheric carbon dioxide has increased from about 280 ppm in 1750 to 383 ppm in 2007 and is increasing at an average rate of 2 ppm pr year. The world's oceans have previously played an important role in sequestering atmospheric carbon dioxide through solubility and the action of phytoplankton. This, and the likely adverse consequences for humans and the biosphere of associated global warming, increases the significance of investigating policy mechanisms for encouraging biosequestration.

Reforestation, Avoided Deforestation and LULUCF

Reforestation and reducing deforestation can increase biosequestration in four ways. Pandani (Richea pandanifolia) near Lake Dobson, Mount Field National Park, Tasmania, Australia

The Intergovernmental Panel on Climate Change (IPCC) estimates that the cutting down of forests is now contributing close to 20 per cent of the overall greenhouse gases entering the atmosphere. Candell and Raupach argue that there are four primary ways in which reforestation and reducing deforestation can increase biosequestration. First, by increasing the volume of existing forest. Second, by increasing the carbon density of existing forests at a stand and landscape scale. Third, by expanding the use of forest products that will sustainably replace fossil-fuel emissions. Fourth, by reducing carbon emissions that are caused from deforestation and degradation. Land clearing reductions, the majority of the time, create biodiversity benefits in a vast expanse of land regions. Concerns, however, arise when the density and area of vegetation increases the grazing pressure could also increase in other areas, causing land degradation.

A recent report by the Australian CSIRO found that forestry and forest-related options are the most significant and most easily achieved carbon sink making up 105 Mt per year CO_2-e or about 75 per cent of the total figure attainable for the Australian state of Queensland from 2010-2050. Among the forestry options, the CSIRO report announced, forestry with the primary aim of carbon storage (called carbon forestry) clearly has the highest attainable carbon storage capacity (77 Mt CO_2-e/yr) and is one of the easiest options to implement compared with biodiversity plantings, pre-1990 eucalypts, post 1990 plantations and managed regrowth. Legal strategies to encourage this form of biosequestration include permanent protection of forests in National Parks or on the World Heritage List, properly funded management and bans on use of rainforest timbers and inefficient uses such as woodchipping old growth forest.

As a result of lobbying by the developing country caucus (or Group of 77) in the United Nations (associated with the United Nations Conference on Environment and Development (UNCED) in Rio de Janeiro, the non-legally binding Forest Principles were established in 1992. These linked the problem of deforestation to third world debt and inadequate technology transfer and stated that the "agreed full incremental cost of achieving benefits associated with forest conservation-should be equitably shared by the international community" (para1(b)). Subsequently, the Group of 77 argued in the 1995 *Intergovernmental Panel on Forests* (IPF) and then the 2001 *Intergovernmental Forum on Forests* (IFF), for affordable access to environmentally sound technologies without the stringency of intellectual property rights; while developed states there rejected demands for a forests fund. The expert group created under the United Nations Forum on Forests

(UNFF) reported in 2004, but in 2007 developed nations again vetoed language in the principles of the final text which might confirm their legal responsibility under international law to supply finance and environmentally sound technologies to the developing world.

Settlement and deforestation surrounding the Brazilian town of Rio Branco are seen here in the striking "herring bone" deforestation patterns that cut through the rainforest. NASA, 2008.

In December 2007, after a two-year debate on a proposal from Papua New Guinea and Costa Rica, state parties to the United Nations Framework Convention on Climate Change (FCCC) agreed to explore ways of reducing emissions from deforestation and to enhance forest carbon stocks in developing nations. The underlying idea is that developing nations should be financially compensated if they succeed in reducing their levels of deforestation (through valuing the carbon that is stored in forests); a concept termed 'avoided deforestation (AD) or, REDD if broadened to include reducing forest degradation. Under the free market model advocated by the countries who have formed the *Coalition of Rainforest Nations*, developing nations with rainforests would sell carbon sink credits under a free market system to Kyoto Protocol Annex I states who have exceeded their emissions allowance. Brazil (the state with the largest area of tropical rainforest) however, opposes including avoided deforestation in a carbon trading mechanism and instead favors creation of a multilateral development assistance fund created from donations by developed states. For REDD to be successful science and regulatory infrastructure related to forests will need to increase so nations may inventory all their forest carbon, show that they can control land use at the local level and prove that their emissions are declining.

NASA Earth Observatory, 2009. Deforestation in Malaysian Borneo.

Subsequent to the initial donor nation response, the UN established REDD Plus, or REDD+, expanding the original program's scope to include increasing forest cover through both reforestation and the planting of new forest cover, as well as promoting sustainable forest resource management.

The United Nations Framework Convention on Climate Change (UNFCCC) Article 4(1)(a) requires all Parties to "develop, periodically update, publish and make available to the Conference of the Parties" as well as "national inventories of anthropogenic emissions by sources" "removals by sinks of all greenhouse gases not controlled by the Montreal Protocol." Under the UNFCCC reporting guidelines, human-induced greenhouse emissions must be reported in six sectors: energy (including stationary energy and transport); industrial processes; solvent and other product use; agriculture; waste; and *land use, land use change and forestry* (LULUCF). The rules governing accounting and reporting of greenhouse gas emissions from LULUCF under the Kyoto Protocol are contained in several decisions of the Conference of Parties under the UNFCCC and LULUCF has been the subject of two major reports by the Intergovernmental Panel on Climate Change (IPCC). The Kyoto Protocol article 3.3 thus requires mandatory LULUCF accounting for afforestation (no forest for last 50 years), reforestation (no forest on 31 December 1989) and deforestation, as well as (in the first commitment period) under article 3.4 voluntary accounting for cropland management, grazing land management, revegetation and forest management (if not already accounted under article 3.3).

Continent of Australia from space. Australia is a major producer of fossil fuels and has significant problems with deforestation.

As an example, the *Australian National Greenhouse Gas Inventory* (NGGI) prepared in compliance with these requirements indicates that the energy sector accounts for 69 per cent of Australia's emissions, agriculture 16 per cent and LULUCF six per cent. Since 1990, however, emissions from the energy sector have increased 35 per cent (stationary energy up 43% and transport up 23%). By comparison, emissions from LULUCF have fallen by 73%. However, questions have been raised by Andrew Macintosh about the veracity of the estimates of emissions from the LULUCF sector because of discrepancies between the Australian Federal and Queensland Governments' land clearing data. Data published by the *Statewide Landcover and Trees Study* (SLATS) in Queensland, for example, show that the total amount of land clearing in Queensland identified under SLATS between 1989/90 and 2000/01 is approximately 50 per cent higher than the amount estimated by the Australian Federal Government's *National Carbon Accounting System* (NCAS) between 1990 and 2001.

Deforestation in Haiti. NASA, 2008.

Satellite imaging has become crucial in obtaining data on levels of deforestation and reforestation. Landsat satellite data, for example, has been used to map tropical deforestation as part of NASA's Landsat *Pathfinder Humid Tropical Deforestation Project*, a collaborative effort among scientists from the University of Maryland, the University of New Hampshire, and NASA's Goddard Space Flight Center. The project yielded deforestation maps for the Amazon Basin, Central Africa, and Southeast Asia for three periods in the 1970s, 1980s, and 1990s.

Enhanced Photosynthesis

Sprekelia formosissima in Tasmania, Australia.

Biosequestration may be enhanced by improving photosynthetic efficiency by modifying RuBisCO genes in plants to increase the catalytic and/or oxygenation activity of that enzyme.

One such research area involves increasing the Earth's proportion of C4 carbon fixation photosynthetic plants. C4 plants represent about 5% of Earth's plant biomass and 1% of its known plant species, but account for around 30% of terrestrial carbon fixation. In leaves of C3 plants, captured photons of solar energy undergo photosynthesis which assimilates carbon into carbohydrates (triosephosphates) in the chloroplasts of the mesophyll cells. The primary CO_2 fixation step is catalysed by ribulose-1,5-bisphosphate carboxylase/oxygenase (Rubisco) which reacts with O2 leading to photorespiration that protects photosynthesis from photoinhibition but wastes 50% of potentially fixed carbon. The C4 photosynthetic pathway, however, concentrates CO_2 at the site of the reaction of Rubisco, thereby reducing the biosequestration-inhibiting photorespiration. A new frontier in crop science consists of attempts to genetically engineer C3 staple food crops (such as wheat, barley, soybeans, potatoes and rice) with the "turbo-charged" photosynthetic apparatus of C4 plants.

Hakea epiglottis, Cape Raoul, Tasman Peninsula, Tasmania, Australia.

Biochar

Biochar (charcoal created by pyrolysis of biomass) is a potent form of longterm (thousands of years) biosequestration of atmosphereic CO_2 derived from investigation of the extremely fertile Terra preta soils of the Amazon Basin. Placing biochar in soils also improves water quality, increases soil fertility, raises agricultural productivity and reduce pressure on old growth forests. As a method of generating bio-energy with carbon storage Rob Flanagan and the EPRIDA biochar company have developed low-tech cooking stoves for developing nations that can burn agricultural wastes such as rice husks and produce 15% by weight of biochar; while BEST Energies in NSW Australia have spent a decade developing an Agrichar technology that can combust 96 tonnes of dry biomass each day, generating 30-40 tonnes of biochar. A parametric study of biosequestration by Malcolm Fowles at the Open University, indicated that to mitigate global warming, policies should encourage displacement of coal with biomass as a power source for baseload electricity generation if the latter's conversion efficiency rose over 30%, otherwise *biosequestering* carbon from biomass as a cheaper mitigation option than geosequestration by CO_2 capture and storage.

Improved Agricultural and Farming Practices

Zero-till farming practices occur where there is much mulching but ploughing is not used, so that the carbon-rich organic matter in soil is not exposed to atmospheric oxygen, or to the leaching and erosion effects of rainfall. Ceasing ploughing has been alleged to encourage more ants to become predators of wood-eating (and CO_2 generating) termites, allows weeds to regenerate soils and helps slow water flows over the land.

Shepherds with their sheep.

Soil holds more carbon than vegetation and atmosphere combined, and most soil lies under grazing land. Holistic Planned Grazing holds tremendous potential in mitigating global warming, while building soil, increasing biodiversity, and reversing desertification. Developed by Allan Savory, it uses fencing and/or herders, to restore grasslands by carefully planning movements of large herds of livestock to mimic the vast herds found in nature where grazing animals are kept concentrated by pack predators and forced to move on after eating, trampling, and manuring an area, returning only after it has fully recovered. This method of grazing seeks to emulate what occurred during the past 40 million years as the expansion of grass-grazer ecosystems built deep, rich grassland soils, sequestering carbon and cooling the planet.

Dedicated biofuel and biosequestration crops, such as switchgrass (panicum virgatum), are also being developed. It requires from 0.97 to 1.34 GJ fossil energy to produce 1 tonne of switchgrass, compared with 1.99 to 2.66 GJ to produce 1 tonne of corn. Given that switchgrass contains approximately 18.8 GJ/ODT of biomass, the energy output-to-input ratio for the crop can be up to 20:1.

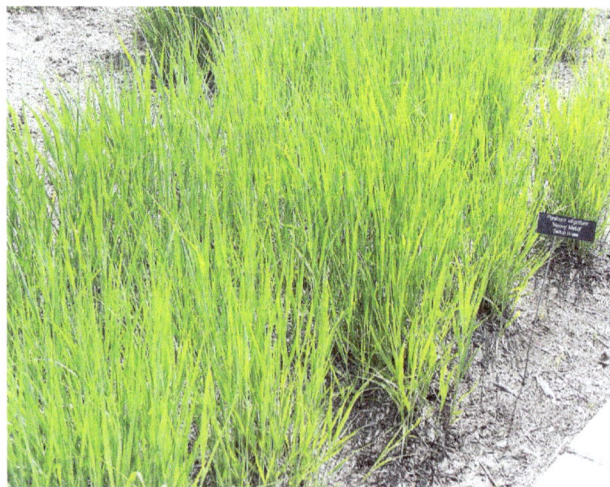

Panicum virgatum switchgrass, valuable in biofuel production, soil conservation and biosequestration

Biosequestration can also be enhanced by farmers choosing crops species that produce large numbers of phytoliths. Phytoliths are microscopic spherical shells of silicon that can store carbon for thousands of years.

Biosequestration and Climate Change Policy

Biosequestration could be critical to climate change mitigation till cleaner forms of power generation are established. The Nesjavellir Geothermal Power Plant in Þingvellir, Iceland

Industries with large amounts of CO_2 emissions (such as the coal industry) are interested in biosequestration as a means of offsetting their greenhouse gas production. In Australia, university researchers are engineering algae to produce biofuels (hydrogen and biodiesel oils) and investigating whether this process can be used to *biosequester* carbon. Algae naturally capture sunlight and use its energy to split water into hydrogen, oxygen and oil which can be extracted. Such clean energy production also can be coupled with desalination using salt-tolerant marine algae to generate fresh water and electricity.

Windturbines D4 (nearest) to D1 on the Thornton Bank

Many new bioenergy (biofuel) technologies, including cellulosic ethanol biorefineries (using stems and branches of most plants including crop residues such as corn stalks, wheat straw and rice straw) are being promoted because they have the added advantage of biosequestration of CO_2. The Garnaut Climate Change Review recommends that a carbon price in a carbon emission trading scheme could include a financial incentive for biosequestration processes. Garnaut recommends the use of algal biosequestration to absorb the constant stream of carbon dioxide emissions from coal-fired electricity generation and metal smelting until renewable forms of energy, such as solar and wind power, become more established contributors to the grid. Garnaut, for example, states:

"Some algal biosequestration processes could absorb emissions from coal-fired electricity generation and metals smelting." The United Nations Collaborative Programme on Reducing Emissions from Deforestation and Forest Degradation in Developing Countries (UN-REDD Programme) is a collaboration between FAO, UNDP and UNEP under which a trust fund established in July 2008 allows donors to pool resources to generate the requisite transfer flow of resources to significantly reduce global emissions from deforestation and forest degradation. The UK government's Stern Review on the economics of climate change argued that curbing deforestation was a "highly cost-effective way of reducing greenhouse gas emissions".

James E. Hansen argues that, "An effective way to achieve drawdown [of carbon dioxide] would be to burn biofuels in power plants and capture the CO_2, with the biofuels derived from agricultural or urban wastes or grown on degraded lands using little or no fossil fuel inputs." Such CO_2 drawdown systems are referred to as Bio-energy with carbon capture and storage, or BECCS. According to a study by Biorecro and the Global CCS Institute, there is currently (as of January 2012) 550 000 tonnes CO_2/year in total BECCS capacity operating, divided between three different facilities.

Under a 2009 agreement, Loy Yang Power and MBD Energy Ltd will build a pilot Fossil fuel power plant at the Latrobe Valley power station in Australia using biosequestration technology in the form of an algal synthesiser system. Captured CO_2 from the waste exhaust flue gases will be injected into circulating waste water to grow oil-rich algae where sunlight and nutrients will produce heavy oil-laden slurry that can make high grade oil for energy, or stock feed. Other commercial demonstration projects involving biosequestration of CO_2 at point of emission have begun in Australia.

Philosophical Basis of Biosequestration

The arguments for biosequestration are often shaped in terms of economic theory, yet there is a well-recognised quality of life dimension to this debate. Biosequestration assists human beings to increase their collective and individual contributions to the essential resources of the biosphere. The policy case for biosequestration overlaps with principles of ecology, sustainability and sustainable development, as well as biosphere, biodiversity and ecosystem protection, environmental ethics, climate ethics and natural conservation.

Barriers to Increased Global Biosequestration

Lassen National Park, Kings Creek, USA.

The Garnaut Climate Change Review notes many barriers to increased global biosequestration. "There must be changes in the accounting regimes for greenhouse gases. Investments are required in research, development and commercialisation of superior approaches to biosequestration. Adjustments are required in the regulation of land use. New institutions will need to be developed to coordinate the interests in utilisation of biosequestration opportunities across small business in rural communities. Special efforts will be required to unlock potential in rural communities in developing countries." Saddler and King have argued that biosequestration and agricultural greenhouse gas emissions should not be handled within a global emissions trading scheme because of difficulties with measuring such emissions, problems in controlling them and the burden that would be placed on numerous small-scale farming operations. Collett likewise maintains that REDD credits (post-facto payments to developing countries for reducing their deforestation rates below an historical or projected reference rate), simply create a complex market approach to this global public health problem that reduces transparency and accountability when targets are not met and will not be as effective as developed nations voluntarily funding countries to keep their rainforests.

The World Rainforest Movement has argued that poor developing countries could be pressured to accept reforestation projects under the Kyoto Protocol's Clean Development Mechanism in order to earn foreign exchange simply to pay off the interest on debt to the World Bank. Tensions also exist over forest management between the sovereignty claims of nations states, arguments about common heritage of mankind and the rights of indigenous peoples and local communities; the Forest Peoples Programme (FPP) arguing the anti-deforestation programs could merely allow financial benefits to flow to national treasuries, privilege would-be corporate forest degraders who manipulate the system by periodically threatening forests, rather than local communities who conserve them. The success of such projects will also depend on the accuracy of the baseline data and the number of countries involved. Further, it has been argued that if biosequestration is to play a significant role in mitigating anthropogenic climate change then coordinated policies should set a goal of achieving global forest cover to its extent prior to the industrial revolution in the 1800s.

It has also been argued that the United Nations mechanism for Reducing Emissions from Deforestation and Forest Degradation (REDD) may increase pressure to convert or modify other ecosystems, especially savannahs and wetlands, for food or biofuel, even though those ecosystems also have high carbon sequestration potential. Globally, for example, peatlands cover only 3% of the land surface but store twice the amount of carbon as all the world's forests, whilst mangrove forests and saltmarshes are examples of relatively low-biomass ecosystems with high levels of productivity and carbon sequestration. Other researchers have argued that REDD is a critical component of an effective global biosequestration strategy that could provide significant benefits, such as the conservation of biodiversity, particularly if it moves away from focusing on protecting forests that are most cost-effective for reducing carbon emissions (such as those in Brazil where agricultural opportunity costs are relatively low, unlike Asia, which has sizeable revenues from oil palm, rubber, rice, and maize). They argue REDD could be varied to allow funding of programs to slow peat degradation in Indonesia and target protection of biodiversity in "hot spot"—areas with high species richness and relatively little remaining forest. Some purchasers, they maintain, of REDD carbon credits, such as multinational corporations or nations, might pay a premium to save imperiled eco-systems or areas with high-profile species.

Carbon Sequestration

Carbon sequestration is the process involved in carbon capture and the long-term storage of atmospheric carbon dioxide. Carbon sequestration involves long-term storage of carbon dioxide or other forms of carbon to mitigate or defer global warming. It has been proposed as a way to slow the atmospheric and marine accumulation of greenhouse gases, which are released by burning fossil fuels.

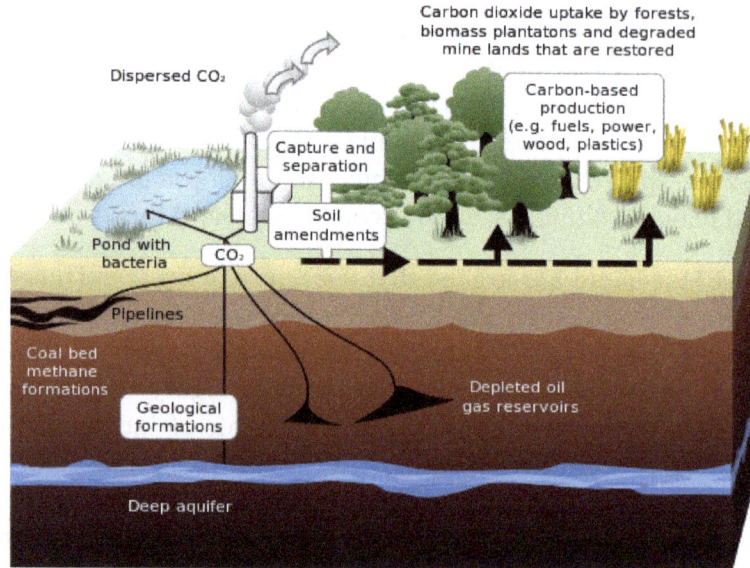

Schematic showing both terrestrial and geological sequestration of carbon dioxide emissions from a coal-fired plant.

Carbon dioxide (CO_2) is naturally captured from the atmosphere through biological, chemical, and physical processes. Artificial processes have been devised to produce similar effects, including large-scale, artificial capture and sequestration of industrially produced CO2 using subsurface saline aquifers, reservoirs, ocean water, aging oil fields, or other carbon sinks.

Description

Carbon sequestration is the process involved in carbon capture and the long-term storage of atmospheric carbon dioxide (CO_2) and may refer specifically to:

- "The process of removing carbon from the atmosphere and depositing it in a reservoir." When carried out deliberately, this may also be referred to as carbon dioxide removal, which is a form of geoengineering.

- Carbon capture and storage, where carbon dioxide is removed from flue gases (e.g., at power stations) before being stored in underground reservoirs.

- Natural biogeochemical cycling of carbon between the atmosphere and reservoirs, such as by chemical weathering of rocks.

Carbon sequestration describes long-term storage of carbon dioxide or other forms of carbon to either mitigate or defer global warming and avoid dangerous climate change. It has been proposed as a way to slow the atmospheric and marine accumulation of greenhouse gases, which are released by burning fossil fuels.

Carbon dioxide is naturally captured from the atmosphere through biological, chemical or physical processes. Some artificial sequestration techniques exploit these natural processes, while some use entirely artificial processes.

Carbon dioxide may be captured as a pure by-product in processes related to petroleum refining or from flue gases from power generation. CO_2 sequestration includes the storage part of carbon capture and storage, which refers to large-scale, artificial capture and sequestration of industrially produced CO_2 using subsurface saline aquifers, reservoirs, ocean water, aging oil fields, or other carbon sinks.

Biological Processes

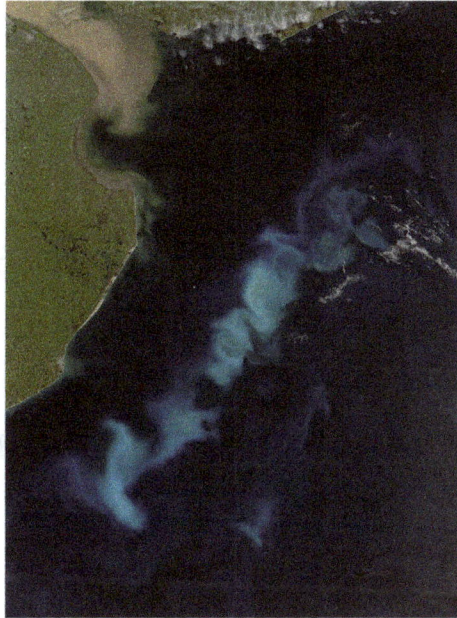

An oceanic phytoplankton bloom in the South Atlantic Ocean, off the coast of Argentina. Encouraging such blooms with iron fertilization could lock up carbon on the seabed.

Biosequestration or carbon sequestration through biological processes affects the global carbon cycle. Examples include major climatic fluctuations, such as the Azolla event, which created the current Arctic climate. Such processes created fossil fuels, as well as clathrate and limestone. By manipulating such processes, geoengineers seek to enhance sequestration.

Peat Production

Peat bogs are a very important carbon store. Peat bogs act as a sink for carbon due to the accumulation of partially decayed biomass that would otherwise continue to decay completely. There is a variance on how much the peatlands act as a carbon sink or carbon source that can be linked to varying climates in different areas of the world and different times of the year. By creating new bogs, or enhancing existing ones, the amount of carbon that is sequestered by bogs would increase.

Forestry

Reforestation is the replanting of trees on marginal crop and pasture lands to incorporate carbon from atmospheric CO_2 into biomass. For this process to succeed the carbon must not return to the atmosphere from mass burning or rotting when the trees die. To this end, land allotted to the trees must not be converted to other uses and management of the frequency of disturbances might be necessary in order to avoid extreme events. Alternatively, the wood from them must itself be sequestered, e.g., via biochar, bio-energy with carbon storage (BECS), landfill or 'stored' by use in e.g. construction. Short of growth in perpetuity, however, reforestation with long-lived trees (>100 years) will sequester carbon for a more graduated release, minimizing impact during the expected carbon crisis of the 21st century.

Urban Forestry

Urban Forestry increases the amount of carbon taken up in cities by adding new tree sites and the sequestration of carbon occurs over the lifetime of the tree. It is generally practiced and maintained on smaller scales, like in cities. The results of urban forestry can have different results depending on the type of vegetation that is being used, so it can function as a sink but can also function as a source of emissions. Along with sequestration by the plants which is difficult to measure but seems to have little effect on the overall amount of carbon dioxide that is uptaken, the vegetation can have indirect effects on carbon by reducing need for energy consumption.

Wetland Restoration

Wetland soil is an important carbon sink; 14.5% of the world's soil carbon is found in wetlands, while only 6% of the world's land is composed of wetlands.

Agriculture

Globally, soils are estimated to contain approximately 1,500 gigatons of organic carbon to 1 m depth, more than the amount in vegetation and the atmosphere.

Modification of agricultural practices is a recognized method of carbon sequestration as soil can act as an effective carbon sink offsetting as much as 20% of 2010 carbon dioxide emissions annually.

Carbon emission reduction methods in agriculture can be grouped into two categories: reducing and/or displacing emissions and enhancing carbon removal. Some of these reductions involve increasing the efficiency of farm operations (e.g. more fuel-efficient equipment) while some involve interruptions in the natural carbon cycle. Also, some effective techniques (such as the elimination of stubble burning) can negatively impact other environmental concerns (increased herbicide use to control weeds not destroyed by burning).

Reducing Emissions

Increasing yields and efficiency generally reduces emissions as well, since more food results from the same or less effort. Techniques include more accurate use of fertilizers, less soil disturbance, better irrigation, and crop strains bred for locally beneficial traits and increased yields.

Replacing more energy intensive farming operations can also reduce emissions. Reduced or no-till farming requires less machine use and burns correspondingly less fuel per acre. However, no-till usually increases use of weed-control chemicals and the residue now left on the soil surface is more likely to release its CO_2 to the atmosphere as it decays, reducing the net carbon reduction.

In practice, most farming operations that incorporate post-harvest crop residues, wastes and by-products back into the soil provide a carbon storage benefit. This is particularly the case for practices such as field burning of stubble - rather than releasing almost all of the stored CO2 to the atmosphere, tillage incorporates the biomass back into the soil.

Enhancing Carbon Removal

All crops absorb CO_2 during growth and release it after harvest. The goal of agricultural carbon removal is to use the crop and its relation to the carbon cycle to permanently sequester carbon within the soil. This is done by selecting farming methods that return biomass to the soil and enhance the conditions in which the carbon within the plants will be reduced to its elemental nature and stored in a stable state. Methods for accomplishing this include:

- Use cover crops such as grasses and weeds as temporary cover between planting seasons

- Concentrate livestock in small paddocks for days at a time so they graze lightly but evenly. This encourages roots to grow deeper into the soil. Stock also till the soil with their hooves, grinding old grass and manures into the soil.

- Cover bare paddocks with hay or dead vegetation. This protects soil from the sun and allows the soil to hold more water and be more attractive to carbon-capturing microbes.

- Restore degraded land, which slows carbon release while returning the land to agriculture or other use.

Agricultural sequestration practices may have positive effects on soil, air, and water quality, be beneficial to wildlife, and expand food production. On degraded croplands, an increase of 1 ton of soil carbon pool may increase crop yield by 20 to 40 kilograms per hectare of wheat, 10 to 20 kg/ha for maize, and 0.5 to 1 kg/ha for cowpeas.

The effects of soil sequestration can be reversed. If the soil is disrupted or tillage practices are abandoned, the soil becomes a net source of greenhouse gases. Typically after 15 to 30 years of sequestration, soil becomes saturated and ceases to absorb carbon. This implies that there is a global limit to the amount of carbon that soil can hold.

Many factors affect the costs of carbon sequestration including soil quality, transaction costs and various externalities such as leakage and unforeseen environmental damage. Because reduction of atmospheric CO_2 is a long-term concern, farmers can be reluctant to adopt more expensive agricultural techniques when there is not a clear crop, soil, or economic benefit. Governments such as Australia and New Zealand are considering allowing farmers to sell carbon credits once they document that they have sufficiently increased soil carbon content.

Ocean-related

Iron Fertilization

Ocean iron fertilization is an example of such a geoengineering technique. Iron fertilization attempts to encourage phytoplankton growth, which removes carbon from the atmosphere for at least a period of time. This technique is controversial due to limited understanding of its complete effects on the marine ecosystem, including side effects and possibly large deviations from expected behavior. Such effects potentially include release of nitrogen oxides, and disruption of the ocean's nutrient balance.

Natural iron fertilisation events (e.g., deposition of iron-rich dust into ocean waters) can enhance carbon sequestration. Sperm whales act as agents of iron fertilisation when they transport iron from the deep ocean to the surface during prey consumption and defecation. Sperm whales have been shown to increase the levels of primary production and carbon export to the deep ocean by depositing iron rich feces into surface waters of the Southern Ocean. The iron rich feces causes phytoplankton to grow and take up more carbon from the atmosphere. When the phytoplankton dies, some of it sinks to the deep ocean and takes the atmospheric carbon with it. By reducing the abundance of sperm whales in the Southern Ocean, whaling has resulted in an extra 200,000 tonnes of carbon remaining in the atmosphere each year.

Urea Fertilization

Ian Jones proposes fertilizing the ocean with urea, a nitrogen rich substance, to encourage phytoplankton growth.

Australian company Ocean Nourishment Corporation (ONC) plans to sink hundreds of tonnes of urea into the ocean to boost CO_2-absorbing phytoplankton growth as a way to combat climate change. In 2007, Sydney-based ONC completed an experiment involving 1 tonne of nitrogen in the Sulu Sea off the Philippines.

Mixing Layers

Encouraging various ocean layers to mix can move nutrients and dissolved gases around, offering avenues for geoengineering. Mixing may be achieved by placing large vertical pipes in the oceans to pump nutrient rich water to the surface, triggering blooms of algae, which store carbon when they grow and export carbon when they die. This produces results somewhat similar to iron fertilization. One side-effect is a short-term rise in CO_2, which limits its attractiveness.

Seaweed

Seaweed grows very fast and can theoretically be harvested and processed to generate biomethane, via Anaerobic Digestion to generate electricity, via Cogeneration/CHP or as a replacement for natural gas. One study suggested that if seaweed farms covered 9% of the ocean they could produce enough biomethane to supply Earth's equivalent demand for fossil fuel energy, remove 53 gigatonnes of CO_2 per year from the atmosphere and sustainably produce 200 kg per year of fish, per person, for 10 billion people. Ideal species for such farming and conversion include Laminaria digitata, Fucus serratus and Saccharina latissima.

Physical Processes

Biochar can be landfilled, used as a soil improver or burned using carbon capture and storage

Biomass-related

Bio-energy with Carbon Capture and Storage (BECCS)

BECCS refers to biomass in power stations and boilers that use carbon capture and storage. The carbon sequestered by the biomass would be captured and stored, thus removing carbon dioxide from the atmosphere.

This technology is sometimes referred to as bio-energy with carbon storage, BECS, though this term can also refer to the carbon sequestration potential in other technologies, such as biochar.

Burial

Burying biomass (such as trees) directly, mimics the natural processes that created fossil fuels. Landfills also represent a physical method of sequestration.

Biochar Burial

Biochar is charcoal created by pyrolysis of biomass waste. The resulting material is added to a landfill or used as a soil improver to create terra preta. Addition of pyrogenic organic carbon (bio-char) is a novel strategy to increase the soil-C stock for the long-term and to mitigate global-warming by offsetting the atmospheric C (up to 9.5 Pg C annually).

In the soil, the carbon is unavailable for oxidation to CO2 and consequential atmospheric release. This is one technique advocated by scientist James Lovelock, creator of the Gaia hypothesis. According to Simon Shackley, "people are talking more about something in the range of one to two billion tonnes a year."

The mechanisms related to biochar are referred to as bio-energy with carbon storage, BECS.

Ocean Storage

If CO_2 were to be injected to the ocean bottom, the pressures would be great enough for CO_2 to be in its liquid phase. The idea behind ocean injection would be to have stable, stationary pools of CO_2

at the ocean floor. The ocean could potentially hold over a thousand billion tons of CO_2. However, this avenue of sequestration isn't being as actively pursued because of concerns about the impact on ocean life, and concerns about its stability.

River mouths bring large quantities of nutrients and dead material from upriver into the ocean as part of the process that eventually produces fossil fuels. Transporting material such as crop waste out to sea and allowing it to sink exploits this idea to increase carbon storage. International regulations on marine dumping may restrict or prevent use of this technique.

Geological Sequestration

Geological sequestration refers to the storage of CO_2 underground in depleted oil and gas reservoirs, saline formations, or deep, un-minable coal beds.

Once CO_2 is captured from a gas or coal-fired power plant, it would be compressed to ≈ 100 bar so that it would be a supercritical fluid. In this fluid form, the CO_2 would be easy to transport via pipeline to the place of storage. The CO_2 would then be injected deep underground, typically around 1 km, where it would be stable for hundreds to millions of years. At these storage conditions, the density of supercritical CO_2 is 600 to 800 kg / m³. For consumers, the cost of electricity from a coal-fired power plant with carbon capture and storage (CCS) is estimated to be 0.01 - 0.05 $ / kWh higher than without CCS. For reference, the average cost of electricity in the US in 2004 was 0.0762 $ / kWh. In other terms, the cost of CCS would be 20 - 70 $/ton of CO_2 captured. The transportation and injection of CO_2 is relatively cheap, with the capture costs accounting for 70 - 80% of CCS costs.

The important parameters in determining a good site for carbon storage are: rock porosity, rock permeability, absence of faults, and geometry of rock layers. The medium in which the CO_2 is to be stored ideally has a high porosity and permeability, such as sandstone or limestone. Sandstone can have a permeability ranging from 1 to 10^{-5} Darcy, and can have a porosity as high as $\approx 30\%$. The porous rock must be capped by a layer of low permeability which acts as a seal, or caprock, for the CO_2. Shale is an example of a very good caprock, with a permeability of 10^{-5} to 10^{-9} Darcy. Once injected, the CO_2 plume will rise via buoyant forces, since it is less dense than its surroundings. Once it encounters a caprock, it will spread laterally until it encounters a gap. If there are fault planes near the injection zone, there is a possibility the CO_2 could migrate along the fault to the surface, leaking into the atmosphere, which would be potentially dangerous to life in the surrounding area. Another danger related to carbon sequestration is induced seismicity. If the injection of CO_2 creates pressures that are too high underground, the formation will fracture, causing an earthquake.

While trapped in a rock formation, CO_2 can be in the supercritical fluid phase or dissolve in groundwater/brine. It can also react with minerals in the geologic formation to precipitate carbonates.

Worldwide storage capacity in oil and gas reservoirs is estimated to be 675 - 900 Gt CO_2, and in un-minable coal seams is estimated to be 15 - 200 Gt CO_2. Deep saline formations have the largest capacity, which is estimated to be 1,000 - 10,000 Gt CO_2. In the US, there is an estimated 160 Gt CO_2 storage capacity.

There are a number of large-scale carbon capture and sequestration projects that have demonstrated the viability and safety of this method of carbon storage, which are summarized here by the Global CCS Institute. The dominant monitoring technique is seismic imaging, where vibrations are generated that propagate through the subsurface. The geologic structure can be imaged from the refracted/reflected waves.

The first large-scale CO_2 sequestration project which began in 1996 is called Sleipner, and is located in the North Sea where Norway's StatoilHydro strips carbon dioxide from natural gas with amine solvents and disposed of this carbon dioxide in a deep saline aquifer. In 2000, a coal-fueled synthetic natural gas plant in Beulah, North Dakota, became the world's first coal-using plant to capture and store carbon dioxide, at the Weyburn-Midale Carbon Dioxide Project.

CO_2 has been used extensively in enhanced crude oil recovery operations in the United States beginning in 1972. There are in excess of 10,000 wells that inject CO_2 in the state of Texas alone. The gas comes in part from anthropogenic sources, but is principally from large naturally occurring geologic formations of CO_2. It is transported to the oil-producing fields through a large network of over 5,000 kilometres (3,100 mi) of CO_2 pipelines. The use of CO_2 for enhanced oil recovery (EOR) methods in heavy oil reservoirs in the Western Canadian Sedimentary Basin (WCSB) has also been proposed. However, transport cost remains an important hurdle. An extensive CO_2 pipeline system does not yet exist in the WCSB. Athabasca oil sands mining that produces CO_2 is hundreds of kilometers north of the subsurface Heavy crude oil reservoirs that could most benefit from CO_2 injection.

Chemical Processes

Developed in the Netherlands, an electrocatalysis by a copper complex helps reduce carbon dioxide to oxalic acid; This conversion uses carbon dioxide as a feedstock to generate oxalic acid.

Mineral Carbonation

Carbon, in the form of CO_2 can be removed from the atmosphere by chemical processes, and stored in stable carbonate mineral forms. This process is known as 'carbon sequestration by mineral carbonation' or mineral sequestration. The process involves reacting carbon dioxide with abundantly available metal oxides–either magnesium oxide (MgO) or calcium oxide (CaO)–to form stable carbonates. These reactions are exothermic and occur naturally (e.g., the weathering of rock over geologic time periods).

$$CaO + CO_2 \rightarrow CaCO_3$$

$$MgO + CO_2 \rightarrow MgCO_3$$

Calcium and magnesium are found in nature typically as calcium and magnesium silicates (such as forsterite and serpentinite) and not as binary oxides. For forsterite and serpentine the reactions are:

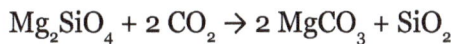

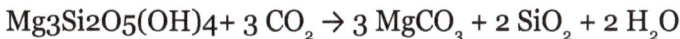

$$Mg_2SiO_4 + 2\ CO_2 \rightarrow 2\ MgCO_3 + SiO_2$$

$$Mg3Si2O5(OH)4 + 3\ CO_2 \rightarrow 3\ MgCO_3 + 2\ SiO_2 + 2\ H_2O$$

The following table lists principal metal oxides of Earth's crust. Theoretically up to 22% of this mineral mass is able to form carbonates.

Earthen Oxide	Percent of Crust	Carbonate	Enthalpy change (kJ/mol)
SiO2	59.71		
Al2O3	15.41		
CaO	4.90	CaCO3	-179
MgO	4.36	MgCO3	-117
Na2O	3.55	Na2CO3	
FeO	3.52	FeCO3	
K2O	2.80	K2CO3	
Fe2O3	2.63	FeCO3	
	21.76	All Carbonates	

These reactions are slightly more favorable at low temperatures. This process occurs naturally over geologic time frames and is responsible for much of the Earth's surface limestone. The reaction rate can be made faster however, by reacting at higher temperatures and/or pressures, although this method requires some additional energy. Alternatively, the mineral could be milled to increase its surface area, and exposed to water and constant abrasion to remove the inert Silica as could be achieved naturally by dumping Olivine in the high energy surf of beaches Experiments suggest the weathering process is reasonably quick (one year) given porous basaltic rocks.

CO_2 naturally reacts with peridotite rock in surface exposures of ophiolites, notably in Oman. It has been suggested that this process can be enhanced to carry out natural mineralisation of CO_2.

Industrial Use

Traditional cement manufacture releases large amounts of carbon dioxide, but newly developed cement types from Novacem can absorb CO_2 from ambient air during hardening. A similar technique was pioneered by TecEco, which has been producing "EcoCement" since 2002.

In Estonia, oil shale ash, generated by power stations could be used as sorbents for CO_2 mineral sequestration. The amount of CO_2 captured averaged 60 to 65% of the carbonaceous CO_2 and 10 to 11% of the total CO_2 emissions.

Chemical Scrubbers

Various carbon dioxide scrubbing processes have been proposed to remove CO_2 from the air, usually using a variant of the Kraft process. Carbon dioxide scrubbing variants exist based on potassium carbonate, which can be used to create liquid fuels, or on sodium hydroxide. These notably include artificial trees proposed by Klaus Lackner to remove carbon dioxide from the atmosphere using chemical scrubbers.

Ocean-related

Basalt Storage

Carbon dioxide sequestration in basalt involves the injecting of CO_2 into deep-sea formations.

The CO_2 first mixes with seawater and then reacts with the basalt, both of which are alkaline-rich elements. This reaction results in the release of Ca^{2+} and Mg^{2+} ions forming stable carbonate minerals.

Underwater basalt offers a good alternative to other forms of oceanic carbon storage because it has a number of trapping measures to ensure added protection against leakage. These measures include "geothermal, sediment, gravitational and hydrate formation." Because CO_2 hydrate is denser than CO_2 in seawater, the risk of leakage is minimal. Injecting the CO_2 at depths greater than 2,700 meters (8,900 ft) ensures that the CO_2 has a greater density than seawater, causing it to sink.

One possible injection site is Juan de Fuca plate. Researchers at the Lamont-Doherty Earth Observatory found that this plate at the western coast of the United States has a possible storage capacity of 208 gigatons. This could cover the entire current U.S. carbon emissions for over 100 years.

This process is undergoing tests as part of the CarbFix project, resulting in 95% of the injected 250 tonnes of CO_2 to solidify into calcite in 2 years, using 25 tonnes of water per tonne of CO_2.

Acid Neutralisation

Carbon dioxide forms carbonic acid when dissolved in water, so ocean acidification is a significant consequence of elevated carbon dioxide levels, and limits the rate at which it can be absorbed into the ocean (the solubility pump). A variety of different bases have been suggested that could neutralize the acid and thus increase CO_2 absorption. For example, adding crushed limestone to oceans enhances the absorption of carbon dioxide. Another approach is to add sodium hydroxide to oceans which is produced by electrolysis of salt water or brine, while eliminating the waste hydrochloric acid by reaction with a volcanic silicate rock such as enstatite, effectively increasing the rate of natural weathering of these rocks to restore ocean pH.

Obstruction

Danger of Leaks

Carbon dioxide may be stored deep underground. At depth, hydrostatic pressure acts to keep it in a liquid state. Reservoir design faults, rock fissures and tectonic processes may act to release the gas stored into the ocean or atmosphere.

Financial Costs

The use of the technology would add an additional 1-5 cents of cost per kilowatt hour, according to estimate made by the Intergovernmental Panel on Climate Change. The financial costs of modern coal technology would nearly double if use of CCS technology were to be required by regulation. The cost of CCS technology differs with the different types of capture technologies being used and with the different sites that it is implemented in, but the costs tend to increase with CCS capture implementation. One study conducted predicted that with new technologies these costs could be lowered but would remain slightly higher than prices without CCS technologies.

Energy Requirements

The energy requirements of sequestration processes may be significant. In one paper, sequestration consumed 25 percent of the plant's rated 600 megawatt output capacity.

After adding CO_2 capture and compression, the capacity of the coal-fired power plant is reduced to 457 MW.

Algae Bioreactor

An algae bioreactor or photobioreactor is used for cultivating algae on purpose to fix CO_2 or produce biomass. Specifically, algae bioreactors can be used to produce fuels such as biodiesel and bioethanol, to generate animal feed, or to reduce pollutants such as NOx and CO_2 in flue gases of power plants. Fundamentally, this kind of bioreactor is based on the photosynthetic reaction which is performed by the chlorophyll-containing algae itself using dissolved carbon dioxide and sunlight energy. The carbon dioxide is dispersed into the reactor fluid to make it accessible for the algae. The bioreactor has to be made out of transparent material.

The algae are photoautotroph organisms which perform oxygenic photosynthesis.

The equation for photosynthesis:

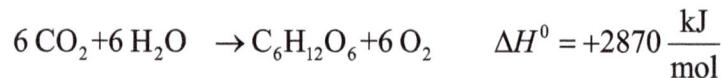

$$6\,CO_2 + 6\,H_2O \;\rightarrow\; C_6H_{12}O_6 + 6\,O_2 \qquad \Delta H^0 = +2870\,\frac{kJ}{mol}$$

Historical Background

Some of the first experiments with the aim of cultivating algae were conducted in 1957 by the "Carnegie Institution" in Washington. In these experiments, the monocellular Chlorella were cultivated by adding CO_2 and some minerals. In the early days, bioreactors were used which were made of glass and later changed to a kind of plastic bag. The goal of all this research has been the cultivation of algae to produce a cheap animal feed.

Frequently used Photo Reactor Types

Nowadays 3 basic types of algae photobioreactors have to be differentiated, but the determining factor is the unifying parameter – the available intensity of sunlight energy.

Plate Photobioreactor

A plate reactor simply consists of vertically arranged or inclined rectangular boxes which are often divided in two parts to effect an agitation of the reactor fluid. Generally these boxes are arranged to a system by linking them. Those connections are also used for making the process of filling/emptying, introduction of gas and transport of nutritive substances, easier. The introduction of the flue gas mostly occurs at the bottom of the box to ensure that the carbon dioxide has enough time to interact with algae in the reactor fluid.

Tubular Photobioreactor

A tubular reactor consists of vertical or horizontal arranged tubes, connected together to a pipe system. The algae-suspended fluid is able to circulate in this tubing. The tubes are generally made out of transparent plastics or borosilicate glass and the constant circulation is kept up by a pump at the end of the system. The introduction of gas takes place at the end/beginning of the tube system. This way of introducing gas causes the problem of deficiency of carbon dioxide, high concentration of oxygen at the end of the unit during the circulation, and bad efficiency.

Bubble Column Photobioreactor

A bubble column photo reactor consists of vertical arranged cylindrical column, made out of transparent material. The introduction of gas takes place at the bottom of the column and causes a turbulent stream to enable an optimum gas exchange. At present these types of reactors are built with a maximum diameter of 20 cm to 30 cm in order to ensure the required supply of sunlight energy.

The biggest problem with the sunlight determined construction is the limited size of the diameter. Feuermann et al. invented a method to collect sunlight with a cone shaped collector and transfer it with some fiberglass cables which are adapted to the reactor in order to enable constructions of a column reactor with wider diameters. - on this scale the energy consumption due to pumps etc. and the CO_2 cost of manufacture may outweigh the CO_2 captured by the reactor.

Industrial Usage

The cultivation of algae in a photobioreactor creates a narrow range of industrial application possibilities. Some power companies already established research facilities with algae photobioreactors to find out how efficient they could be in reducing CO_2 emissions, which are contained in flue gas, and how much biomass will be produced. Algae biomass has many uses and can be sold to generate additional income. The saved emission volume can bring an income too, by selling emission credits to other power companies.

The utilisation of algae as food is very common in East Asian regions. Most of the species contain only a fraction of usable proteins and carbohydrates, and a lot of minerals and trace elements. Generally, the consumption of algae should be minimal because of the high iodine content. For example, if someone has hyperthyroidism it could be very dangerous for their health. Likewise, many species of diatomaceous algae produce compounds unsafe for humans.

The algae, especially some species which contain over 50 percent oil and a lot of carbohydrates, can be used for producing biodiesel and bioethanol by extracting and refining the fractions. This point is very interesting, because the algae biomass is generated 30 times faster than some agricultural biomass, which is commonly used for producing biodiesel.

Climate Change Mitigation

Climate change mitigation consists of actions to limit the magnitude or rate of long-term climate change. Climate change mitigation generally involves reductions in human (anthropogenic) emissions of greenhouse gases (GHGs). Mitigation may also be achieved by increasing the capacity of carbon sinks, e.g., through reforestation. Mitigation policies can substantially reduce the risks associated with human-induced global warming.

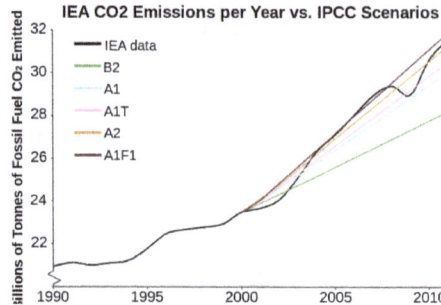

Fossil fuel related CO_2 emissions compared to five of IPCC's emissions scenarios. The dips are related to global recessions. Data from IPCC SRES scenarios; Data spreadsheet included with International Energy Agency's "CO_2 Emissions from Fuel Combustion 2010 – Highlights"; and Supplemental IEA data.

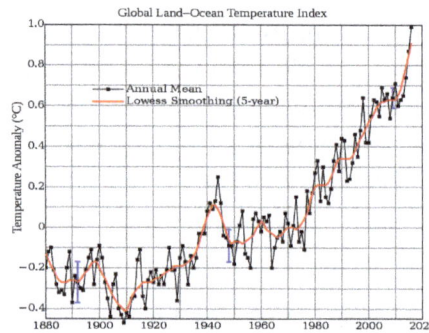

Global mean surface temperature change from 1880 to 2016, relative to the 1951–1980 mean. The black line is the global annual mean and the red line is the five-year lowess smooth. The blue uncertainty bars show a 95% confidence limit. Global dimming, from sulfate aerosol air pollution, between 1950 and 1980 is believed to have mitigated global warming somewhat.

According to the IPCC's 2014 assessment report, "Mitigation is a public good; climate change is a case of the 'tragedy of the commons'. Effective climate change mitigation will not be achieved if each agent (individual, institution or country) acts independently in its own selfish interest, suggesting the need for collective action. Some adaptation actions, on the other hand, have characteristics of a private good as benefits of actions may accrue more directly to the individuals, regions, or countries that undertake them, at least in the short term. Nevertheless, financing such adaptive activities remains an issue, particularly for poor individuals and countries."

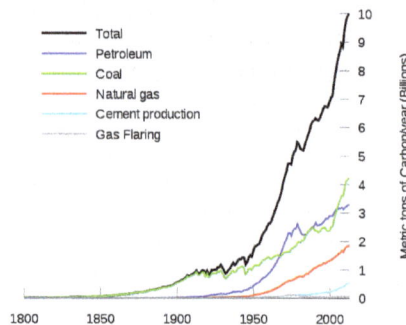

Global carbon dioxide emissions from human activities 1800–2007.

Annual Greenhouse Gas Emissions by Sector

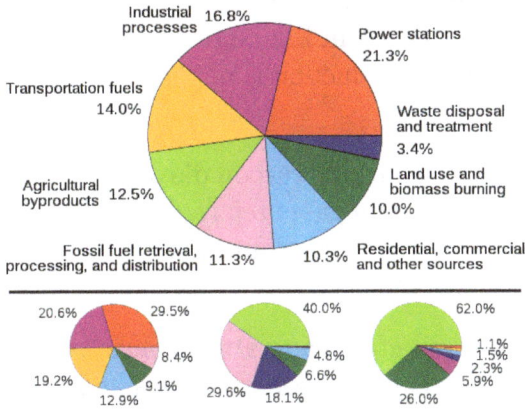

Industrial processes 16.8%
Power stations 21.3%
Transportation fuels 14.0%
Waste disposal and treatment 3.4%
Agricultural byproducts 12.5%
Land use and biomass burning 10.0%
Fossil fuel retrieval, processing, and distribution 11.3%
Residential, commercial and other sources 10.3%

20.6% 29.5%
8.4%
19.2% 9.1%
12.9%

40.0%
4.8%
6.6%
29.6% 18.1%

62.0%
1.1%
1.5%
2.3%
5.9%
26.0%

Greenhouse gas emissions by sector.

Global public support for energy sources

"Please indicate whether you strongly support, somewhat support, somewhat oppose, or strongly oppose each way of producing energy"

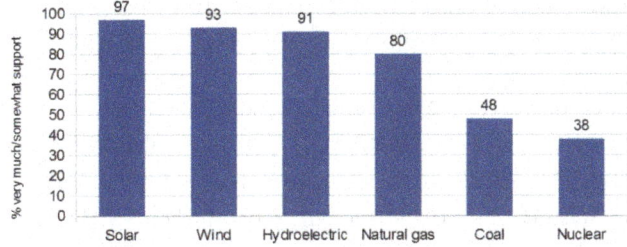

% very much/somewhat support

Solar 97
Wind 93
Hydroelectric 91
Natural gas 80
Coal 48
Nuclear 38

Global public support for energy sources, based on a survey by Ipsos (2011).

Examples of mitigation include phasing out fossil fuels by switching to low-carbon energy sources, such as renewable and nuclear energy, and expanding forests and other "sinks" to remove greater amounts of carbon dioxide from the atmosphere. Energy efficiency may also play a role, for example, through improving the insulation of buildings. Another approach to climate change mitigation is climate engineering.

Most countries are parties to the United Nations Framework Convention on Climate Change (UNFCCC). The ultimate objective of the UNFCCC is to stabilize atmospheric concentrations of GHGs at a level that would prevent dangerous human interference of the climate system. Scientific analysis can provide information on the impacts of climate change, but deciding which impacts are dangerous requires value judgments.

In 2010, Parties to the UNFCCC agreed that future global warming should be limited to below 2.0 °C (3.6 °F) relative to the pre-industrial level. With the Paris Agreement of 2015 this was confirmed, but was revised with a new target laying down "parties will do the best" to achieve warming below 1.5 °C. The current trajectory of global greenhouse gas emissions does not appear to be consistent with limiting global warming to below 1.5 or 2 °C. Other mitigation policies have been proposed, some of which are more stringent or modest than the 2 °C limit.

Greenhouse Gas Concentrations and Stabilization

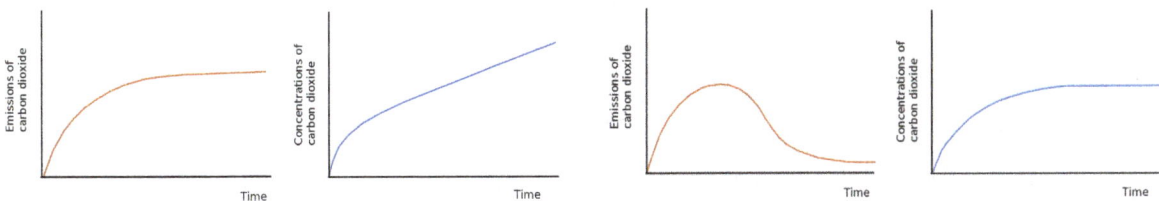

Stabilizing CO_2 emissions at their present level would not stabilize its concentration in the atmosphere.

Stabilizing the atmospheric concentration of CO_2 at a constant level would require emissions to be effectively eliminated.

One of the issues often discussed in relation to climate change mitigation is the stabilization of greenhouse gas concentrations in the atmosphere. The United Nations Framework Convention on

Climate Change (UNFCCC) has the ultimate objective of preventing "dangerous" anthropogenic (i.e., human) interference of the climate system. As is stated in Article 2 of the Convention, this requires that greenhouse gas (GHG) concentrations are stabilized in the atmosphere at a level where ecosystems can adapt naturally to climate change, food production is not threatened, and economic development can proceed in a sustainable fashion.

There are a number of anthropogenic greenhouse gases. These include carbon dioxide (chemical formula: CO_2), methane (CH_4), nitrous oxide (N_2O), and a group of gases referred to as halocarbons. The emissions reductions necessary to stabilize the atmospheric concentrations of these gases varies. CO_2 is the most important of the anthropogenic greenhouse gases.

There is a difference between stabilizing CO_2 emissions and stabilizing atmospheric concentrations of CO_2. Stabilizing emissions of CO_2 at current levels would not lead to a stabilization in the atmospheric concentration of CO_2. In fact, stabilizing emissions at current levels would result in the atmospheric concentration of CO_2 continuing to rise over the 21st century and beyond.

The reason for this is that human activities are adding CO_2 to the atmosphere faster than natural processes can remove it. This is analogous to a flow of water into a bathtub. So long as the tap runs water (analogous to the emission of carbon dioxide) into the tub faster than water escapes through the plughole (the natural removal of carbon dioxide from the atmosphere), then the level of water in the tub (analogous to the concentration of carbon dioxide in the atmosphere) will continue to rise.

According to some studies, stabilizing atmospheric CO_2 concentrations would require anthropogenic CO_2 emissions to be reduced by 80% relative to the peak emissions level. An 80% reduction in emissions would stabilize CO_2 concentrations for around a century, but even greater reductions would be required beyond this. Other research has found that, after leaving room for emissions for food production for 9 billion people and to keep the global temperature rise below 2 °C, emissions from energy production and transport will have to peak almost immediately in the developed world and decline at ca. 10% per annum until zero emissions are reached around 2030. In developing countries energy and transport emissions would have to peak by 2025 and then decline similarly.

Stabilizing the atmospheric concentration of the other greenhouse gasses humans emit also depends on how fast their emissions are added to the atmosphere, and how fast the GHGs are removed. Stabilization for these gases is described in the later section on non-CO_2 GHGs.

Projections

Projections of future greenhouse gas emissions are highly uncertain. In the absence of policies to mitigate climate change, GHG emissions could rise significantly over the 21st century.

Numerous assessments have considered how atmospheric GHG concentrations could be stabilized. The lower the desired stabilization level, the sooner global GHG emissions must peak and decline. GHG concentrations are unlikely to stabilize this century without major policy changes.

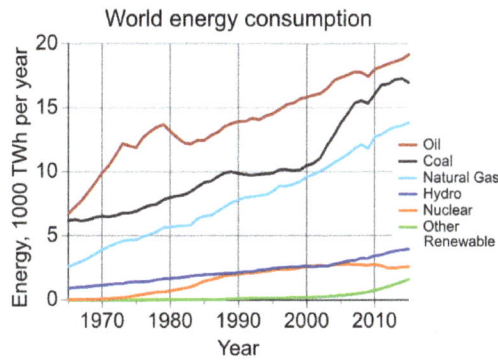

The above image shows rate of world energy usage per day, from 1970 to 2010. Every fossil fuel source has increased in large amounts between 1970 and 2010, dominating all other energy sources. Hydroelectricity has increased at a slow steady rate over this same period, nuclear entered a period of rapid growth between 1970 and 1990 before leveling off. Other renewables, between 2000 and 2010 have, having started from a low usage rate, began to enter into a period of rapid growth. 1000 TWh=1 PWh.

Energy Consumption by Power Source

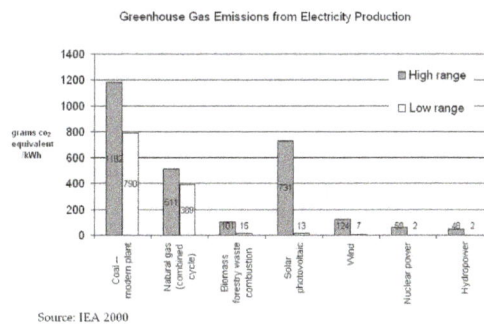

"Hydropower-Internalised Costs and Externalised Benefits"; Frans H. Koch; International Energy Agency (IEA)-Implementing Agreement for Hydropower Technologies and Programmes; 2000.

To create lasting climate change mitigation, the replacement of high carbon emission intensity power sources, such as conventional fossil fuels—oil, coal and natural gas—with low-carbon power sources is required. Fossil fuels supply humanity with the vast majority of our energy demands, and at a growing rate. In 2012 the IEA noted that coal accounted for half the increased energy use of the prior decade, growing faster than all renewable energy sources. Both hydroelectricity and nuclear power together provide the majority of the generated low-carbon power fraction of global total power consumption.

Fuel type	Average total global power consumption in TW		
	1980	2004	2006
Oil	4.38	5.58	5.74
Gas	1.80	3.45	3.61
Coal	2.34	3.87	4.27

Hydroelectric	0.60	0.93	1.00
Nuclear power	0.25	0.91	0.93
Geothermal, wind, solar energy, wood	0.02	0.13	0.16
Total	**9.48**	**15.0**	**15.8**

Change and use of energy, by source, in units of (PWh) in that year.				
	Fossil	**Nuclear**	**All renewables**	**Total**
1990	83.374	6.113	13.082	102.569
2000	94.493	7.857	15.337	117.687
2008	117.076	8.283	18.492	143.851
Change 2000–2008	22.583	0.426	3.155	26.164

Methods and Means

Assessments often suggest that GHG emissions can be reduced using a portfolio of low-carbon technologies. At the core of most proposals is the reduction of greenhouse gas (GHG) emissions through reducing energy waste and switching to low-carbon power sources of energy. As the cost of reducing GHG emissions in the electricity sector appears to be lower than in other sectors, such as in the transportation sector, the electricity sector may deliver the largest proportional carbon reductions under an economically efficient climate policy.

"Economic tools can be useful in designing climate change mitigation policies." "While the limitations of economics and social welfare analysis, including cost–benefit analysis, are widely documented, economics nevertheless provides useful tools for assessing the pros and cons of taking, or not taking, action on climate change mitigation, as well as of adaptation measures, in achieving competing societal goals. Understanding these pros and cons can help in making policy decisions on climate change mitigation and can influence the actions taken by countries, institutions and individuals."

Other frequently discussed means include energy conservation, increasing fuel economy in automobiles (which includes the use of electric hybrids), charging plug-in hybrids and electric cars by low-carbon electricity, making individual-lifestyle changes (e.g., cycling instead of driving), and changing business practices. Many fossil fuel driven vehicles can be converted to use electricity, the US has the potential to supply electricity for 73% of light duty vehicles (LDV), using overnight charging. The US average CO_2 emissions for a battery-electric car is 180 grams per mile vs 430 grams per mile for a gasoline car. The emissions would be displaced away from street level, where they have "high human-health implications. Increased use of electricity "generation for meeting the future transportation load is primarily fossil-fuel based", mostly natural gas, followed by coal, but could also be met through nuclear, tidal, hydroelectric and other sources.

A range of energy technologies may contribute to climate change mitigation. These include nuclear power and renewable energy sources such as biomass, hydroelectricity, wind power, solar power, geothermal power, ocean energy, and; the use of carbon sinks, and carbon capture and storage. For example, Pacala and Socolow of Princeton have proposed a 15 part program to reduce CO_2 emissions by 1 billion metric tons per year – or 25 billion tons over the 50-year period using today's technologies as a type of Global warming game.

Another consideration is how future socio-economic development proceeds. Development choices (or "pathways") can lead differences in GHG emissions. Political and social attitudes may affect how easy or difficult it is to implement effective policies to reduce emissions.

Alternative Energy Sources

Renewable Energy

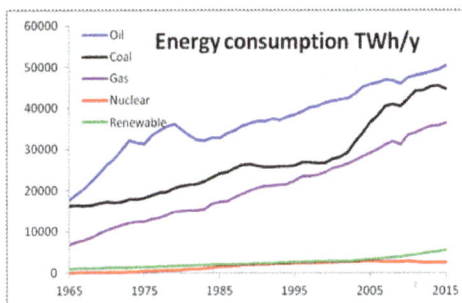

The worldwide growth of renewable energy is shown by the green line

The 22,500 MW nameplate capacity Three Gorges Dam in the Peoples Republic of China, the largest hydroelectric power station in the world.

The Shepherds Flat Wind Farm is an 845 megawatt (MW) nameplate capacity, wind farm in the US state of Oregon, each turbine is a nameplate 2 or 2.5 MW electricity generator.

The 150 MW Andasol solar power station is a commercial parabolic trough solar thermal power plant, located in Spain. The Andasol plant uses tanks of molten salt to store solar energy so that it can continue generating electricity for 7.5 hours after the sun has stopped shining.

Renewable energy flows involve natural phenomena such as sunlight, wind, rain, tides, plant growth, and geothermal heat, as the International Energy Agency explains:

Renewable energy is derived from natural processes that are replenished constantly. In its various forms, it derives directly from the sun, or from heat generated deep within the earth. Included in the definition is electricity and heat generated from solar, wind, ocean, hydropower, biomass, geo-thermal resources, and biofuels and hydrogen derived from renewable resources.

Climate change concerns and the need to reduce carbon emissions are driving increasing growth in the renewable energy industries. Low-carbon renewable energy replaces conventional fossil fuels in three main areas: power generation, hot water/ space heating, and transport fuels. In 2011, the share of renewables in electricity generation worldwide grew for the fourth year in a row to 20.2%. Based on REN21's 2014 report, renewables contributed 19% to supply global energy consumption. This energy consumption is divided as 9% coming from burning biomass, 4.2% as heat energy (non-biomass), 3.8% hydro electricity and 2% as electricity from wind, solar, geothermal, and biomass thermal power plants.

Renewable energy use has grown much faster than anyone anticipated. The Intergovernmental Panel on Climate Change (IPCC) has said that there are few fundamental technological limits to integrating a portfolio of renewable energy technologies to meet most of total global energy demand. At the national level, at least 30 nations around the world already have renewable energy contributing more than 20% of energy supply.

As of 2012, renewable energy accounts for almost half of new electricity capacity installed and costs are continuing to fall. Public policy and political leadership helps to "level the playing field" and drive the wider acceptance of renewable energy technologies. As of 2011, 118 countries have targets for their own renewable energy futures, and have enacted wide-ranging public policies to promote renewables. Leading renewable energy companies include BrightSource Energy, First Solar, Gamesa, GE Energy, Goldwind, Sinovel, Suntech, Trina Solar, Vestas and Yingli.

The incentive to use 100% renewable energy has been created by global warming and other ecological as well as economic concerns. Mark Z. Jacobson says producing all new energy with wind power, solar power, and hydropower by 2030 is feasible and existing energy supply arrangements could be replaced by 2050. Barriers to implementing the renewable energy plan are seen to be "primarily social and political, not technological or economic". Jacobson says that energy costs with a wind, solar, water system should be similar to today's energy costs. According to a 2011 projection by the (IEA)International Energy Agency, solar power generators may produce most of the world's electricity within 50 years, dramatically reducing harmful greenhouse gas emissions. Critics of the "100% renewable energy" approach include Vaclav Smil and James E. Hansen. Smil and Hansen are concerned about the variable output of solar and wind power, NIMBYism, and a lack of infrastructure.

Economic analysts expect market gains for renewable energy (and efficient energy use) following the 2011 Japanese nuclear accidents. In his 2012 State of the Union address, President Barack Obama restated his commitment to renewable energy and mentioned the long-standing Interior Department commitment to permit 10,000 MW of renewable energy projects on public land in 2012. Globally, there are an estimated 3 million direct jobs in renewable energy industries, with about half of them in the biofuels industry.

Some countries, with favorable geography, geology and weather well suited to an economical exploitation of renewable energy sources, already get most of their electricity from renewables, including from geothermal energy in Iceland (100 percent), and Hydroelectric power in Brazil (85 percent), Austria (62 percent), New Zealand (65 percent), and Sweden (54 percent). Renewable power generators are spread across many countries, with wind power providing a significant share of electricity in some regional areas: for example, 14 percent in the US state of Iowa, 40 percent in the northern German state of Schleswig-Holstein, and 20 percent in Denmark. Solar water heating makes an important and growing contribution in many countries, most notably in China, which now has 70 percent of the global total (180 GWth). Worldwide, total installed solar water heating systems meet a portion of the water heating needs of over 70 million households. The use of biomass for heating continues to grow as well. In Sweden, national use of biomass energy has surpassed that of oil. Direct geothermal heating is also growing rapidly. Renewable biofuels for transportation, such as ethanol fuel and biodiesel, have contributed to a significant decline in oil consumption in the United States since 2006. The 93 billion liters of biofuels produced worldwide in 2009 displaced the equivalent of an estimated 68 billion liters of gasoline, equal to about 5 percent of world gasoline production.

Some of the world's largest solar power stations: Ivanpah (CSP) and Topaz (PV), both in California

Nuclear Power

Since about 2001 the term "nuclear renaissance" has been used to refer to a possible nuclear power industry revival, driven by rising fossil fuel prices and new concerns about meeting greenhouse gas emission limits. However, in March 2011 the Fukushima nuclear disaster in Japan and associated shutdowns at other nuclear facilities raised questions among some commentators over the future of nuclear power. Platts has reported that "the crisis at Japan's Fukushima nuclear plants has prompted leading energy-consuming countries to review the safety of their existing reactors and cast doubt on the speed and scale of planned expansions around the world".

The World Nuclear Association has reported that nuclear electricity generation in 2012 was at its lowest level since 1999. Several previous international studies and assessments, suggested that as part of the portfolio of other low-carbon energy technologies, nuclear power will continue to play a role in reducing greenhouse gas emissions. Historically, nuclear power usage is estimated to have prevented the atmospheric emission of 64 gigatonnes of CO_2-equivalent as of 2013. Public concerns about nuclear power include the fate of spent nuclear fuel, nuclear accidents, security risks, nuclear proliferation, and a concern that nuclear power plants are very expensive. Of these concerns, nuclear accidents and disposal of long-lived radioactive fuel/"waste" have probably had the greatest public impact worldwide. Although generally unaware of it, both of these glaring public concerns are greatly diminished by present passive safety designs, the experimentally proven, "melt-down proof" EBR-II, future molten salt reactors, and the use of conventional and more advanced fuel/"waste" pyroprocessing, with the latter recycling or reprocessing not presently being commonplace as it is often considered to be cheaper to use a once-through nuclear fuel cycle in many countries, depending on the varying levels of intrinsic value given by a society in reducing the long-lived waste in their country, with France doing a considerable amount of reprocessing when compared to the US.

Blue Cherenkov light being produced near the core of the Fission powered Advanced Test Reactor

Nuclear power, with a 10.6% share of world electricity production as of 2013, is second only to hydroelectricity as the largest source of low-carbon power. Over 400 reactors generate electricity in 31 countries.

A Yale University review published in the Journal of Industrial Ecology analyzing CO_2 life cycle assessment(LCA) emissions from nuclear power(Light water reactors) determined that: "The collective LCA literature indicates that life cycle GHG emissions from nuclear power are only a fraction of traditional fossil sources and comparable to renewable technologies." While some have raised uncertainty surrounding the future GHG emissions of nuclear power as a result of an extreme potential decline in uranium ore grade without a corresponding increase in the efficiency of enrichment methods. In a scenario analysis of future global nuclear development, as it could be effected by a decreasing global uranium market of average ore grade, the analysis determined that depending on conditions, median life cycle nuclear power GHG emissions could be between 9 and 110 g CO_2-eq/kWh by 2050, with the latter high figure being derived from a "worst-case scenario" that is not "considered very robust" by the authors of the paper, as the "ore grade" in the scenario is lower than the uranium concentration in many lignite coal ashes.

Although this future analyses primarily deals with extrapolations for present Generation II reactor technology, the same paper also summarizes the literature on "FBRs"/Fast Breeder Reactors, of which two are in operation as of 2014 with the newest being the BN-800, for these reactors it states that the "median life cycle GHG emissions ... [are] similar to or lower than [present light water reactors] LWRs and purports to consume little or no uranium ore.

In their 2014 report, the IPCC comparison of energy sources global warming potential per unit of electricity generated, which notably included albedo effects, mirror the median emission value derived from the Warner and Heath Yale meta-analysis for the more common non-breeding Light water reactors, a CO2-equivalent value of 12 g CO_2-eq/kWh, which is the lowest global warming forcing of all baseload power sources, with comparable low carbon power baseload sources, such as hydropower and biomass, producing substantially more global warming forcing 24 and 230 g CO_2-eq/kWh respectively.

In 2014, Brookings Institution published *The Net Benefits of Low and No-Carbon Electricity Technologies* which states, after performing an energy and emissions cost analysis, that "The net benefits of new nuclear, hydro, and natural gas combined cycle plants far outweigh the net benefits of new wind or solar plants", with the most cost effective low carbon power technology being determined to be nuclear power.

During his presidential campaign, Barack Obama stated, "Nuclear power represents more than 70% of our noncarbon generated electricity. It is unlikely that we can meet our aggressive climate goals if we eliminate nuclear power as an option."

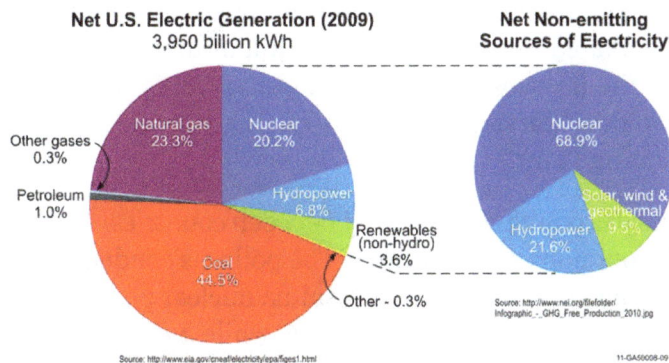

This graph illustrates nuclear power is the United States's largest contributor of non-greenhouse-gas-emitting electric power generation, comprising nearly three-quarters of the non-emitting sources.

Analysis in 2015 by Professor and Chair of Environmental Sustainability Barry W. Brook and his colleagues on the topic of replacing fossil fuels entirely, from the electric grid of the world, has determined that at the historically modest and proven-rate at which nuclear energy was added to and replaced fossil fuels in France and Sweden during each nation's building programs in the 1980s, within 10 years nuclear energy could displace or remove fossil fuels from the electric grid completely, "allow[ing] the world to meet the most stringent greenhouse-gas mitigation targets.". In a similar analysis, Brook had earlier determined that 50% of all global energy, that is not solely electricity, but transportation synfuels etc. could be generated within approximately 30 years, if the global nuclear fission build rate was identical to each of these nation's already proven decadal rates(in units of installed nameplate capacity, GW per year, per unit of global GDP(GW/year/$).

This is in contrast to the completely conceptual paper-studies for a *100% renewable energy* world, which would require an orders of magnitude more costly global investment per year, an investment rate that has no historical precedent, having never been attempted due to its prohibitive cost, and with far greater land area that would be required to be devoted to the wind, wave and solar projects, along with the inherent assumption that humanity will use less, and not more, energy in the future. As Brook notes the "principal limitations on nuclear fission are not technical, economic or fuel-related, but are instead linked to complex issues of societal acceptance, fiscal and political inertia, and inadequate critical evaluation of the real-world constraints facing [the other] low-carbon alternatives."

Nuclear power may be uncompetitive compared with fossil fuel energy sources in countries without a carbon tax program, and in comparison to a fossil fuel plant of the same power output, nuclear power plants take a longer amount of time to construct.

Two new, first of their kind, EPR reactors under construction in Finland and France have been delayed and are running over-budget. However learning from experience, two further EPR reactors under construction in China are on, and ahead, of schedule respectively. As of 2013, according to the IAEA and the European Nuclear Society, worldwide there were 68 civil nuclear power reactors under construction in 15 countries. China has 29 of these nuclear power reactors under construction, as of 2013, with plans to build many more, while in the US the licenses of almost half its reactors have been extended to 60 years, and plans to build another dozen are under serious consideration. There are also a considerable number of new reactors being built in South Korea, India, and Russia. At least 100 older and smaller reactors will "most probably be closed over the next 10–15 years". This is probable only if one does not factor in the ongoing Light Water Reactor Sustainability Program, created to permit the extension of the life span of the USA's 104 nuclear reactors to 60 years. The licenses of almost half of the USA's reactors have been extended to 60 years as of 2008. Two new "passive safety" AP1000 reactors are, as of 2013, being constructed at Vogtle Electric Generating Plant.

Public opinion about nuclear power varies widely between countries. A poll by Gallup International (2011) assessed public opinion in 47 countries. The poll was conducted following a tsunami and earthquake which caused an accident at the Fukushima nuclear power plant in Japan. 49% stated that they held favourable views about nuclear energy, while 43% held an unfavourable view. Another global survey by Ipsos (2011) assessed public opinion in 24 countries. Respondents to this survey showed a clear preference for renewable energy sources over coal and nuclear energy (refer to graph opposite). Ipsos (2012) found that solar and wind were viewed by the public as being more environmentally friendly and more viable long-term energy sources relative to nuclear power and natural gas. However, solar and wind were viewed as being less reliable relative to nuclear power and natural gas. In 2012 a poll done in the UK found that 63% of those surveyed support nuclear power, and with opposition to nuclear power at 11%. In Germany, strong anti-nuclear sentiment led to eight of the seventeen operating reactors being permanently shut down following the March 2011 Fukushima nuclear disaster.

Nuclear fusion research, in the form of the International Thermonuclear Experimental Reactor is underway. Fusion powered electricity generation was initially believed to be readily achievable, as fission power had been. However, the extreme requirements for continuous reactions and plasma containment led to projections being extended by several decades. In 2010, more than 60 years after the first attempts, commercial power production was still believed to be unlikely before 2050. Although rather than an either, or, issue economical fusion-fission hybrid reactors could be built before any attempt at this more demanding commercial "pure-fusion reactor"/DEMO reactor takes place.

Coal to Gas Fuel Switching

Most mitigation proposals imply—rather than directly state—an eventual reduction in global fossil fuel production. Also proposed are direct quotas on global fossil fuel production.

Natural gas emits far fewer greenhouse gases (i.e. CO_2 and methane—CH_4) than coal when burned at power plants, but evidence has been emerging that this benefit could be completely negated by methane leakage at gas drilling fields and other points in the supply chain.

A study performed by the Environmental Protection Agency (EPA) and the Gas Research Institute (GRI) in 1997 sought to discover whether the reduction in carbon dioxide emissions from increased natural gas (predominantly methane) use would be offset by a possible increased level of methane emissions from sources such as leaks and emissions. The study concluded that the reduction in emissions from increased natural gas use outweighs the detrimental effects of increased methane emissions. More recent peer-reviewed studies have challenged the findings of this study, with researchers from the National Oceanic and Atmospheric Administration (NOAA) reconfirming findings of high rates of methane (CH_4) leakage from natural gas fields.

A 2011 study by noted climate research scientist, Tom Wigley, found that while carbon dioxide (CO_2) emissions from fossil fuel combustion may be reduced by using natural gas rather than coal to produce energy, it also found that additional methane (CH_4) from leakage adds to the radiative forcing of the climate system, offsetting the reduction in CO_2 forcing that accompanies the transition from coal to gas. The study looked at methane leakage from coal mining; changes in radiative forcing due to changes in the emissions of sulfur dioxide and carbonaceous aerosols; and differences in the efficiency of electricity production between coal- and gas-fired power generation. On balance, these factors more than offset the reduction in warming due to reduced CO_2 emissions. When gas replaces coal there is additional warming out to 2,050 with an assumed leakage rate of 0%, and out to 2,140 if the leakage rate is as high as 10%. The overall effects on global-mean temperature over the 21st century, however, are small. Petron et al. (2013) and Alvarez et al. (2012) note that estimated that leakage from gas infrastructure is likely to be underestimated. These studies indicate that the exploitation of natural gas as a "cleaner" fuel is questionable. A 2014 meta-study of 20 years of natural gas technical literature shows that methane emissions are consistently underestimated but on a 100-year scale, the climate benefits of coal to gas fuel switching are likely larger than the negative effects of natural gas leakage.

Heat Pump

A heat pump is a device that provides heat energy from a source of heat to a destination called a "heat sink". Heat pumps are designed to move thermal energy opposite to the direction of spontaneous heat flow by absorbing heat from a cold space and releasing it to a warmer one. A heat pump uses some amount of external power to accomplish the work of transferring energy from the heat source to the heat sink.

While air conditioners and freezers are familiar examples of heat pumps, the term "heat pump" is more general and applies to many HVAC (heating, ventilating, and air conditioning) devices used for space heating or space cooling. When a heat pump is used for heating, it employs the same basic refrigeration-type cycle used by an air conditioner or a refrigerator, but in the opposite direction—releasing heat into the conditioned space rather than the surrounding environment. In this use, heat pumps generally draw heat from the cooler external air or from the ground. In heating mode, heat pumps are three to four times more efficient in their use of electric power than simple electrical resistance heaters.

Outside unit of an air-source heat pump

It has been concluded that heat pumps are the single technology that could reduce the greenhouse gas emissions of households better than every other technology that is available on the market. With a market share of 30% and (potentially) clean electricity, heat pumps could reduce global CO_2 emissions by 8% annually. Using ground source heat pumps could reduce around 60% of the primary energy demand and 90% of CO_2 emissions in Europe in 2050 and make handling high shares of renewable energy easier. Using surplus renewable energy in heat pumps is regarded as the most effective household means to reduce global warming and fossil fuel depletion.

With significant amounts of fossil fuel used in electricity production, demands on the electrical grid also generate greenhouse gases. Without a high share of low-carbon electricity, a domestic heat pump will produce more carbon emissions than using natural gas.

Fossil Fuel Phase-out: Carbon Neutral and Negative Fuels

3,500–4,000 environmental activists blocking a coal mine in Germany to limit climate change (Ende Gelände 2016)

Fossil fuel may be phased-out with carbon neutral and carbon negative pipeline and transportation fuels created with power to gas and gas to liquids technologies. Carbon dioxide from fossil fuel flue gas can be used to produce plastic lumber allowing carbon negative reforestation.

Demand Side Management

Energy Efficiency and Conservation

A spiral-type integrated compact fluorescent lamp, use has grown among North American consumers since its introduction in the mid-1990s.

Efficient energy use, sometimes simply called "energy efficiency", is the goal of efforts to reduce the amount of energy required to provide products and services. For example, insulating a home allows a building to use less heating and cooling energy to achieve and maintain a comfortable temperature. Installing fluorescent lights or natural skylights reduces the amount of energy required to attain the same level of illumination compared to using traditional incandescent light bulbs. Compact fluorescent lights use two-thirds less energy and may last 6 to 10 times longer than incandescent lights.

Energy efficiency has proved to be a cost-effective strategy for building economies without necessarily growing energy consumption. For example, the state of California began implementing energy-efficiency measures in the mid-1970s, including building code and appliance standards with strict efficiency requirements. During the following years, California's energy consumption has remained approximately flat on a per capita basis while national US consumption doubled. As part of its strategy, California implemented a "loading order" for new energy resources that puts energy efficiency first, renewable electricity supplies second, and new fossil-fired power plants last.

Energy conservation is broader than energy efficiency in that it encompasses using less energy to achieve a lesser energy demanding service, for example through behavioral change, as well as encompassing energy efficiency. Examples of conservation without efficiency improvements would be heating a room less in winter, driving less, or working in a less brightly lit room. As with other definitions, the boundary between efficient energy use and energy conservation can be fuzzy, but both are important in environmental and economic terms. This is especially the case when actions are directed at the saving of fossil fuels.

Reducing energy use is seen as a key solution to the problem of reducing greenhouse gas emissions. According to the International Energy Agency, improved energy efficiency in buildings, industrial processes and transportation could reduce the world's energy needs in 2050 by one third, and help control global emissions of greenhouse gases.

Demand Side Switching Sources

Fuel switching on the demand side refers to changing the type of fuel used to satisfy a need for an energy service. To meet deep decarbonization goals, like the 80% reduction by 2050 goal being discussed in California and the European Union, many primary energy changes are needed. Energy efficiency alone may not be sufficient to meet these goals, switching fuels used on the demand side will help lower carbon emissions. Progressively coal, oil and eventually natural gas for space and water heating in buildings will need to be reduced. For an equivalent amount of heat, burning natural gas produces about 45 per cent less carbon dioxide than burning coal. There are various ways in which this could happen, and different strategies will likely make sense in different locations. While the system efficiency of a gas furnace may be higher than the combination of natural gas power plant and electric heat, the combination of the same natural gas power plant and an electric heat pump has lower emissions per unit of heat delivered in all but the coldest climates. This is possible because of the very efficient coefficient of performance of heat pumps.

At the beginning of this century 70% of all electricity was generated by fossil fuels, and as carbon free sources eventually make up half of the generation mix, replacing gas or oil furnaces and water heaters with electric ones will have a climate benefit. In areas like Norway, Brazil and Quebec that have abundant hydroelectricity, electric heat and hot water is common.

The economics of switching the demand side from fossil fuels to electricity for heating, will depend on the price of fuels vs electricity and the relative prices of the equipment. The EIA Annual Energy Outlook 2014 suggests that domestic gas prices will rise faster than electricity prices which will encourage electrification in the coming decades. Electrifying heating loads may also provide a flexible resource that can participate in demand response. Since thermostatically controlled loads have inherent energy storage, electrification of heating could provide a valuable resource to integrate variable renewable resources into the grid.

Alternatives to electrification, include decarbonizing pipeline gas through power to gas, biogas, or other carbon neutral fuels. A 2015 study by Energy+Environmental Economics shows that a hybrid approach of decarbonizing pipeline gas, electrification, and energy efficiency can meet carbon reduction goals at a similar cost as only electrification and energy efficiency in Southern California.

Demand Side Grid Management

Expanding intermittent electrical sources such as wind power, creates a growing problem balancing grid fluctuations. Some of the plans include building pumped storage or continental super grids costing billions of dollars. However instead of building for more power,there are a variety of ways to affect the size and timing of electricity demand on the consumer side. Designing for reduced demands on a smaller power grid is more efficient and economic than having extra generation and transmission for intermittentcy, power failures and peak demands. Having these abilities is one of the chief aims of a smart grid.

Time of use metering is a common way to motivate electricity users to reduce their peak load consumption. For instance, running dishwashers and laundry at night after the peak has passed, reduces electricity costs.

Dynamic demand plans have devices passively shut off when stress is sensed on the electrical grid.

This method may work very well with thermostats, when power on the grid sags a small amount, a low power temperature setting is automatically selected reducing the load on the grid. For instance millions of refrigerators reduce their consumption when clouds pass over solar installations. Consumers would need to have a smart meter in order for the utility to calculate credits.

Demand response devices could receive all sorts of messages from the grid. The message could be a request to use a low power mode similar to dynamic demand, to shut off entirely during a sudden failure on the grid, or notifications about the current and expected prices for power. This would allow electric cars to recharge at the least expensive rates independent of the time of day. The vehicle-to-grid suggestion would use a car's battery or fuel cell to supply the grid temporarily.

Lifestyle and Behavior

The IPCC Fifth Assessment Report emphasises that behaviour, lifestyle and cultural change have a high mitigation potential in some sectors, particularly when complementing technological and structural change. In general, higher consumption lifestyles have a greater environmental impact. Overall, food accounts for the largest share of consumption-based GHG emissions with nearly 20% of the global carbon footprint, followed by housing, mobility, services, manufactured products, and construction. Food and services are more significant in poor countries, while mobility and manufactured goods are more significant in rich countries.

Dietary Change

A 2014 study into the real-life diets of British people estimates their greenhouse gas contributions (CO_2eq) to be: 7.19 kg/day for high meat-eaters through to 3.81 kg/day for vegetarians and 2.89 kg/day for vegans. The widespread adoption of a vegetarian diet could cut food-related greenhouse gas emissions by 63% by 2050. China introduced new dietary guidelines in 2016 which aim to cut meat consumption by 50% and thereby reduce greenhouse gas emissions by 1 billion tonnes by 2030. A 2016 study concluded that taxes on meat and milk could simultaneously result in reduced greenhouse gas emissions and healthier diets. The study analyzed surcharges of 40% on beef and 20% on milk and suggests that an optimum plan would reduce emissions by 1 billion tonnes per year.

Sinks and Negative Emissions

A carbon sink is a natural or artificial reservoir that accumulates and stores some carbon-containing chemical compound for an indefinite period, such as a growing forest. A negative carbon dioxide emission on the other hand is a permanent removal of carbon dioxide out of the atmosphere, such as directly capturing carbon dioxide in the atmosphere and storing it in geologic formations underground.

The Antarctic Climate and Ecosystems Cooperative Research Centre (ACE-CRC) notes that one third of humankind's annual emissions of CO_2 are absorbed by the oceans. However, this also leads to ocean acidification, with potentially significant impacts on marine life. Acidification lowers the level of carbonate ions available for calcifying organisms to form their shells. These organisms include plankton species that contribute to the foundation of the Southern Ocean food web. However acidification may impact on a broad range of other physiological and ecological processes, such as fish respiration, larval development and changes in the solubility of both nutrients and toxins.

Reforestation and Afforestation

Transferring land rights to indigenous inhabitants is argued to efficiently conserve forests. Regrowth of forests on abandoned farmland restores more forest than that lost to deforestation.

Almost 20 percent (8 $GtCO_2$/year) of total greenhouse-gas emissions were from deforestation in 2007. It is estimated that avoided deforestation reduces CO_2 emissions at a rate of 1 tonne of CO_2 per \$1–5 in opportunity costs from lost agriculture. Reforestation and afforestation, where there was previously no forest, could save at least another 1 $GtCO_2$/year, at an estimated cost of \$5–15/tCO_2.

Transferring rights over land from public domain to its indigenous inhabitants is argued to be a cost effective strategy to conserve forests. This includes the protection of such rights entitled in existing laws, such as India's Forest Rights Act. The transferring of such rights in China, perhaps the largest land reform in modern times, has been argued to have increased forest cover. In Brazil, forested areas given tenure to indigenous groups have even lower rates of clearing than national parks. A 2016 report concludes that modest investments in indigenous land rights will generate economic, social, and environmental returns for the communities involved and for climate protection. The report quantifies the economic value of securing such rights, with a focus on the Amazon region.

With increased intensive agriculture and urbanization, there is an increase in the amount of abandoned farmland. By some estimates, for every half a hectare of original old-growth forest cut down, more than 20 hectares of new secondary forests are growing, even though they do not have the same biodiversity as the original forests and original forests store 60% more carbon than these new secondary forests. According to a study in *Science*, promoting regrowth on abandoned farmland could offset years of carbon emissions.

It appears that attempts to reduce atmospheric carbon through biomass absorption coupled with sequestration can succeed only if combined with early and far-reaching reduction of current greenhouse gas generation by all means available. Absent such reduction, pursuit of biomass absorption plus sequestration as a late-regret measure would require prohibitively massive use of resources in order to succeed.

Avoided Desertification

Restoring grasslands store CO_2 from the air into plant material. Grazing livestock, usually not left to wander, would eat the grass and would minimize any grass growth. However, grass left alone would eventually grow to cover its own growing buds, preventing them from photosynthesizing and the dying plant would stay in place. A method proposed to restore grasslands uses fences with many small paddocks and moving herds from one paddock to another after a day a two in order to mimick natural grazers and allowing the grass to grow optimally. Additionally, when part of leaf matter is consumed by a herding animal, a corresponding amount of root matter is sloughed off too as it would not be able to sustain the previous amount of root matter and while most of the lost root matter would rot and enter the atmosphere, part of the carbon is sequestered into the soil. It is estimated that increasing the carbon content of the soils in the world's 3.5 billion hectares of agricultural grassland by 1% would offset nearly 12 years of CO_2 emissions. Allan Savory, as part of holistic management, claims that while large herds are often blamed for desertification, prehistoric lands supported large or larger herds and areas where herds were removed in the United States are still desertifying.

Managed grazing methods are argued to be able to restore grasslands, thereby significantly decreasing atmospheric CO_2 levels.

Carbon Capture and Storage

Carbon capture and storage (CCS) is a method to mitigate climate change by capturing carbon dioxide (CO_2) from large point sources such as power plants and subsequently storing it away safely instead of releasing it into the atmosphere. The Intergovernmental Panel on Climate Change says CCS could contribute between 10% and 55% of the cumulative worldwide carbon-mitigation effort over the next 90 years. The International Energy Agency says CCS is "the most important single new technology for CO_2 savings" in power generation and industry. Though it requires up to 40% more energy to run a CCS coal power plant than a regular coal plant, CCS could potentially capture about 90% of all the carbon emitted by the plant. Norway, which first began storing CO_2, has cut its emissions by almost a million tons a year, or about 3% of the country's 1990 levels. As of late 2011, the total CO_2 storage capacity of all 14 projects in operation or under construction is over 33 million tonnes a year. This is broadly equivalent to preventing the emissions from more than six million cars from entering the atmosphere each year.

Negative Carbon Dioxide Emissions

Creating negative carbon dioxide emissions literally removes carbon from the atmosphere. Examples are direct air capture, biochar, bio-energy with carbon capture and storage and enhanced weathering technologies. These processes are sometimes considered as variations of sinks or mitigation, and sometimes as geoengineering.

In combination with other mitigation measures, sinks in combination with negative carbon emissions are considered crucial for meeting the 350 ppm target, and even the less conservative 450 ppm target.

Geoengineering

Geoengineering is seen by some as an alternative to mitigation and adaptation, but by others as an entirely separate response to climate change. In a literature assessment, Barker *et al.* (2007) described geoengineering as a type of mitigation policy. IPCC (2007) concluded that geoengineering options, such as ocean fertilization to remove CO_2 from the atmosphere, remained largely unproven. It was judged that reliable cost estimates for geoengineering had not yet been published.

Chapter 28 of the National Academy of Sciences report *Policy Implications of Greenhouse Warming: Mitigation, Adaptation, and the Science Base* (1992) defined geoengineering as "options that would involve large-scale engineering of our environment in order to combat or counteract the effects of changes in atmospheric chemistry." They evaluated a range of options to try to give preliminary answers to two questions: can these options work and could they be carried out with a reasonable cost. They also sought to encourage discussion of a third question — what adverse side effects might there be. The following types of option were examined: reforestation, increasing ocean absorption of carbon dioxide (carbon sequestration) and screening out some sunlight. NAS also argued "Engineered countermeasures need to be evaluated but should not be implemented without broad understanding of the direct effects and the potential side effects, the ethical issues, and the risks.". In July 2011 a report by the United States Government Accountability Office on geoengineering found that "[c]limate engineering technologies do not now offer a viable response to global climate change."

Carbon Dioxide Removal

Carbon dioxide removal has been proposed as a method of reducing the amount of radiative forcing. A variety of means of artificially capturing and storing carbon, as well as of enhancing natural sequestration processes, are being explored. The main natural process is photosynthesis by plants and single-celled organisms. Artificial processes vary, and concerns have been expressed about the long-term effects of some of these processes.

It is notable that the availability of cheap energy and appropriate sites for geological storage of carbon may make carbon dioxide air capture viable commercially. It is, however, generally expected that carbon dioxide air capture may be uneconomic when compared to carbon capture and storage from major sources — in particular, fossil fuel powered power stations, refineries, etc. In such cases, costs of energy produced will grow significantly. However, captured CO_2 can be used to force more crude oil out of oil fields, as Statoil and Shell have made plans to do. CO_2 can also be used in commercial greenhouses, giving an opportunity to kick-start the technology. Some attempts have been made to use algae to capture smokestack emissions, notably the GreenFuel Technologies Corporation, who have now shut down operations.

Solar Radiation Management

The main purpose of solar radiation management seek to reflect sunlight and thus reduce global warming. The ability of stratospheric sulfate aerosols to create a global dimming effect has made them a possible candidate for use in climate engineering projects.

Non-CO$_2$ Greenhouse Gases

CO$_2$ is not the only GHG relevant to mitigation, and governments have acted to regulate the emissions of other GHGs emitted by human activities (anthropogenic GHGs). The emissions caps agreed to by most developed countries under the Kyoto Protocol regulate the emissions of almost all the anthropogenic GHGs. These gases are CO$_2$, methane (CH$_4$), nitrous oxide (N$_2$O), the hydrofluorocarbons (HFC), perfluorocarbons (PFC), and sulfur hexafluoride (SF$_6$).

Stabilizing the atmospheric concentrations of the different anthropogenic GHGs requires an understanding of their different physical properties. Stabilization depends both on how quickly GHGs are added to the atmosphere and how fast they are removed. The rate of removal is measured by the atmospheric lifetime of the GHG in question. Here, the lifetime is defined as the time required for a given perturbation of the GHG in the atmosphere to be reduced to 37% of its initial amount. Methane has a relatively short atmospheric lifetime of about 12 years, while N$_2$O's lifetime is about 110 years. For methane, a reduction of about 30% below current emission levels would lead to a stabilization in its atmospheric concentration, while for N$_2$O, an emissions reduction of more than 50% would be required.

Methane is a significantly more potent greenhouse gas than carbon dioxide in the amount of heat it can trap, especially in the short term. Burning one molecule of methane generates one molecule of carbon dioxide, indicating there may be no net benefit in using gas as a fuel source. Reducing the amount of waste methane produced in the first place and moving away from use of gas as a fuel source will have a greater beneficial impact, as might other approaches to productive use of otherwise-wasted methane. In terms of prevention, vaccines are being developed in Australia to reduce the significant global warming contributions from methane released by livestock via flatulence and eructation.

Another physical property of the anthropogenic GHGs relevant to mitigation is the different abilities of the gases to trap heat (in the form of infrared radiation). Some gases are more effective at trapping heat than others, e.g., SF$_6$ is 22,200 times more effective a GHG than CO$_2$ on a per-kilogram basis. A measure for this physical property is the global warming potential (GWP), and is used in the Kyoto Protocol.

Although not designed for this purpose, the Montreal Protocol has probably benefited climate change mitigation efforts. The Montreal Protocol is an international treaty that has successfully reduced emissions of ozone-depleting substances (for example, CFCs), which are also greenhouse gases.

By Sector

Transport

Transportation emissions account for roughly 1/4 of emissions worldwide, and are even more important in terms of impact in developed nations especially in North America and Australia. Many citizens of countries like the United States and Canada who drive personal cars often, see well over half of their climate change impact stemming from the emissions produced from their cars. Modes of mass transportation such as bus, light rail (metro, subway, etc.), and long-distance rail are far and away the most energy-efficient means of motorized transportation for passengers, able to use in many cases over twenty times less energy per person-distance than a personal automobile. Modern energy-efficient technologies, such as plug-in hybrid electric vehicles and carbon-neutral synthetic gasoline & Jet fuel may also help to reduce the consumption of petroleum, land use changes and emissions of carbon dioxide. Utilizing rail transport, especially electric rail, over the far less efficient air transport and truck transport significantly reduces emissions. With the use of electric trains and cars in transportation there is the opportunity to run them with low-carbon power, producing far fewer emissions.

The Tesla Roadster emits no tailpipe emissions, uses lithium ion batteries to achieve 220 mi (350 km) per charge, while also capable of going 0–60 in under 4 seconds.

Bicycles have almost no carbon footprint compared to cars, and canal transport may represent a positive option for certain types of freight in the 21st century.

Urban Planning

Effective urban planning to reduce sprawl aims to decrease Vehicle Miles Travelled (VMT), lowering emissions from transportation. Personal cars are extremely inefficient at moving passengers, while public transport and bicycles are many times more efficient (as is the simplest form of human transportation, walking). All of these are encouraged by urban/community planning and are an effective way to reduce greenhouse gas emissions. Between 1982 and 1997, the amount of land consumed for urban development in the United States increased by 47 percent while the nation's population grew by only 17 percent. Inefficient land use development practices have increased infrastructure costs as well as the amount of energy needed for transportation, community services, and buildings.

At the same time, a growing number of citizens and government officials have begun advocating a smarter approach to land use planning. These smart growth practices include compact community development, multiple transportation choices, mixed land uses, and practices to conserve green space. These programs offer environmental, economic, and quality-of-life benefits; and they also serve to reduce energy usage and greenhouse gas emissions.

Approaches such as New Urbanism and Transit-oriented development seek to reduce distances travelled, especially by private vehicles, encourage public transit and make walking and cycling more attractive options. This is achieved through "medium-density", mixed-use planning and the concentration of housing within walking distance of town centers and transport nodes.

Smarter growth land use policies have both a direct and indirect effect on energy consuming behavior. For example, transportation energy usage, the number one user of petroleum fuels, could be significantly reduced through more compact and mixed use land development patterns, which in turn could be served by a greater variety of non-automotive based transportation choices.

Building Design

Emissions from housing are substantial, and government-supported energy efficiency programmes can make a difference.

For institutions of higher learning in the United States, greenhouse gas emissions depend primarily on total area of buildings and secondarily on climate. If climate is not taken into account, annual greenhouse gas emissions due to energy consumed on campuses plus purchased electricity can be estimated with the formula, $E=aS^b$, where a =0.001621 metric tonnes of CO_2 equivalent/square foot or 0.0241 metric tonnes of CO_2 equivalent/square meter and b = 1.1354.

New buildings can be constructed using passive solar building design, low-energy building, or zero-energy building techniques, using renewable heat sources. Existing buildings can be made more efficient through the use of insulation, high-efficiency appliances (particularly hot water heaters and furnaces), double- or triple-glazed gas-filled windows, external window shades, and building orientation and siting. Renewable heat sources such as shallow geothermal and passive solar energy reduce the amount of greenhouse gasses emitted. In addition to designing buildings which are more energy-efficient to heat, it is possible to design buildings that are more energy-efficient to cool by using lighter-coloured, more reflective materials in the development of urban areas (e.g. by painting roofs white) and planting trees. This saves energy because it cools buildings and reduces the urban heat island effect thus reducing the use of air conditioning.

Agriculture

According to the EPA, agricultural soil management practices can lead to production and emission of nitrous oxide (N_2O), a major greenhouse gas and air pollutant. Activities that can contribute to N_2O emissions include fertilizer usage, irrigation and tillage. The management of soils accounts for over half of the emissions from the Agriculture sector. Cattle livestocks account for one third of emissions, through methane emissions. Manure management and rice cultivation also produce gaseous emissions.

Methods that significantly enhance carbon sequestration in soil include no-till farming, residue mulching, cover cropping, and crop rotation, all of which are more widely used in organic farming than in conventional farming. Because only 5% of US farmland currently uses no-till and residue mulching, there is a large potential for carbon sequestration.

A 2015 study found that farming can deplete soil carbon and render soil incapable of supporting life; however, the study also showed that conservation farming can protect carbon in soils, and repair damage over time.

The farming practise of cover crops has been recognized as climate-smart agriculture by the White House.

Societal Controls

Another method being examined is to make carbon a new currency by introducing tradeable "personal carbon credits". The idea being it will encourage and motivate individuals to reduce their 'carbon footprint' by the way they live. Each citizen will receive a free annual quota of carbon that they can use to travel, buy food, and go about their business. It has been suggested that by using this concept it could actually solve two problems; pollution and poverty, old age pensioners will actually be better off because they fly less often, so they can cash in their quota at the end of the year to pay heating bills and so forth.

Population

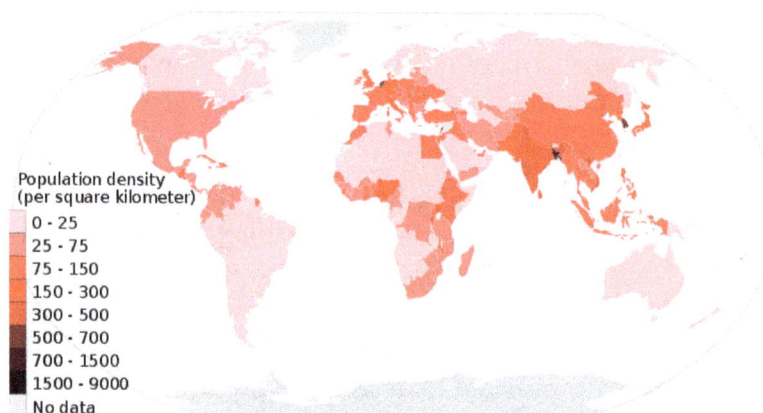

Population density by country

Various organizations promote population control as a means for mitigating global warming. Proposed measures include improving access to family planning and reproductive health care and information, reducing natalistic politics, public education about the consequences of continued population growth, and improving access of women to education and economic opportunities.

Population control efforts are impeded by there being somewhat of a taboo in some countries against considering any such efforts. Also, various religions discourage or prohibit some or all forms of birth control.

Population size has a different per capita effect on global warming in different countries, since the per capita production of anthropogenic greenhouse gases varies greatly by country.

Costs and Benefits

Costs

The Stern Review proposes stabilising the concentration of greenhouse-gas emissions in the atmosphere at a maximum of 550ppm CO_2e by 2050. The Review estimates that this would mean cutting total greenhouse-gas emissions to three quarters of 2007 levels. The Review further estimates that the cost of these cuts would be in the range −1.0 to +3.5% of World GDP, (i.e. GWP), with an average estimate of approximately 1%. Stern has since revised his estimate to 2% of GWP. For comparison, the Gross World Product (GWP) at PPP was estimated at $74.5 trillion in 2010, thus 2% is approximately $1.5 trillion. The Review emphasises that these costs are contingent on steady reductions in the cost of low-carbon technologies. Mitigation costs will also vary according to how and when emissions are cut: early, well-planned action will minimise the costs.

One way of estimating the cost of reducing emissions is by considering the likely costs of potential technological and output changes. Policy makers can compare the marginal abatement costs of different methods to assess the cost and amount of possible abatement over time. The marginal abatement costs of the various measures will differ by country, by sector, and over time.

Benefits

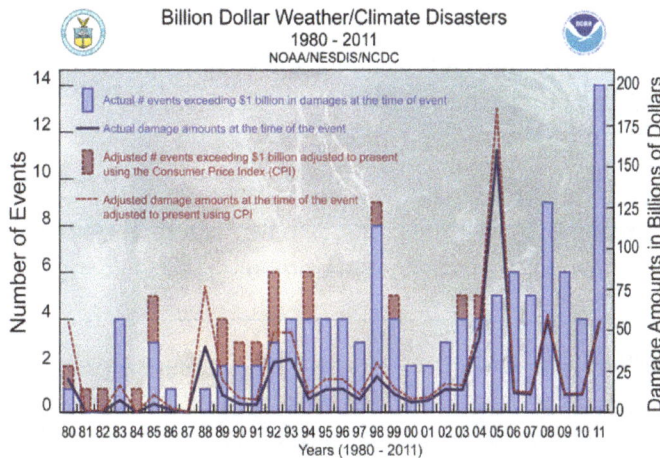

Total extreme weather cost and number of events costing more than $1 billion in the United States from 1980 to 2011

Yohe *et al.* (2007) assessed the literature on sustainability and climate change. With high confidence, they suggested that up to the year 2050, an effort to cap greenhouse gas (GHG) emissions at 550 ppm would benefit developing countries significantly. This was judged to be especially the case when combined with enhanced adaptation. By 2100, however, it was still judged likely that there would be significant effects of global warming. This was judged to be the case even with aggressive mitigation and significantly enhanced adaptive capacity.

Sharing

One of the aspects of mitigation is how to share the costs and benefits of mitigation policies. There is no scientific consensus over how to share these costs and benefits (Toth *et al.*, 2001). In terms of the politics of mitigation, the UNFCCC's ultimate objective is to stabilize concentrations of GHG in the atmosphere at a level that would prevent "dangerous" climate change (Rogner *et al.*, 2007).

GHG emissions are an important correlate of wealth, at least at present (Banuri *et al.*, 1996, pp. 91–92). Wealth, as measured by per capita income (i.e., income per head of population), varies widely between different countries. Activities of the poor that involve emissions of GHGs are often associated with basic needs, such as heating to stay tolerably warm. In richer countries, emissions tend to be associated with things like cars, central heating, etc. The impacts of cutting emissions could therefore have different impacts on human welfare according to wealth.

Distributing Emissions Abatement Costs

There have been different proposals on how to allocate responsibility for cutting emissions (Banuri *et al.*, 1996, pp. 103–105):

- Egalitarianism: this system interprets the problem as one where each person has equal rights to a global resource, i.e., polluting the atmosphere.

- Basic needs: this system would have emissions allocated according to basic needs, as defined according to a minimum level of consumption. Consumption above basic needs would require countries to buy more emission rights. From this viewpoint, developing countries would need to be at least as well off under an emissions control regime as they would be outside the regime.

- Proportionality and polluter-pays principle: Proportionality reflects the ancient Aristotelian principle that people should receive in proportion to what they put in, and pay in proportion to the damages they cause. This has a potential relationship with the "polluter-pays principle", which can be interpreted in a number of ways:

  - *Historical responsibilities*: this asserts that allocation of emission rights should be based on patterns of past emissions. Two-thirds of the stock of GHGs in the atmosphere at present is due to the past actions of developed countries (Goldemberg *et al.*, 1996, p. 29).

  - *Comparable burdens and ability to pay*: with this approach, countries would reduce emissions based on comparable burdens and their ability to take on the costs of reduction. Ways to assess burdens include monetary costs per head of population, as well as other, more complex measures, like the UNDP's Human Development Index.

  - *Willingness to pay*: with this approach, countries take on emission reductions based on their ability to pay along with how much they benefit from reducing their emissions.

Specific Proposals

- Ad hoc: Lashof (1992) and Cline (1992), for example, suggested that allocations based partly on GNP could be a way of sharing the burdens of emission reductions. This is because GNP and economic activity are partially tied to carbon emissions.

- Equal per capita entitlements: this is the most widely cited method of distributing abatement costs, and is derived from egalitarianism (Banuri *et al.*, 1996, pp. 106–107). This approach can be divided into two categories. In the first category, emissions are allocated according to national population. In the second category, emissions are allocated in a way that attempts to account for historical (cumulative) emissions.

- Status quo: with this approach, historical emissions are ignored, and current emission levels are taken as a status quo right to emit. An analogy for this approach can be made with fisheries, which is a common, limited resource. The analogy would be with the atmosphere, which can be viewed as an exhaustible natural resource. In international law, one state recognized the long-established use of another state's use of the fisheries resource. It was also recognized by the state that part of the other state's economy was dependent on that resource.

Governmental and Intergovernmental Action

Many countries, both developing and developed, are aiming to use cleaner technologies. Use of these technologies aids mitigation and could result in substantial reductions in CO_2 emissions. Policies include targets for emissions reductions, increased use of renewable energy, and increased energy efficiency. It is often argued that the results of climate change are more damaging in poor nations, where infrastructures are weak and few social services exist. The Commitment to Development Index is one attempt to analyze rich country policies taken to reduce their disproportionate use of the global commons. Countries do well if their greenhouse gas emissions are falling, if their gas taxes are high, if they do not subsidize the fishing industry, if they have a low fossil fuel rate per capita, and if they control imports of illegally cut tropical timber.

Kyoto Protocol

The main current international agreement on combating climate change is the Kyoto Protocol, which came into force on 16 February 2005. The Kyoto Protocol is an amendment to the United Nations Framework Convention on Climate Change (UNFCCC). Countries that have ratified this protocol have committed to reduce their emissions of carbon dioxide and five other greenhouse gases, or engage in emissions trading if they maintain or increase emissions of these gases.

Temperature Targets

Actions to mitigate climate change are sometimes based on the goal of achieving a particular temperature target. One of the targets that has been suggested is to limit the future increase in global mean temperature (global warming) to below 2 °C, relative to the pre-industrial level. The 2 °C target was adopted in 2010 by Parties to the United Nations Framework Convention on Climate Change. Most countries of the world are Parties to the UNFCCC. The target had been adopted in 1996 by the European Union Council.

Feasibility of 2 °C

Temperatures have increased by 0.8 °C compared to the pre-industrial level, and another 0.5–0.7 °C is already committed. The 2 °C rise is typically associated in climate models with a carbon dioxide equivalent concentration of 400–500 ppm by volume; the current (January 2015) level of carbon dioxide alone is 400 ppm by volume, and rising at 1–3 ppm annually. Hence, to avoid a very likely breach of the 2 °C target, CO_2 levels would have to be stabilised very soon; this is generally regarded as unlikely, based on current programs in place to date. The importance of change is illustrated by the fact that world economic energy efficiency is improving at only half the rate of world economic growth.

Views in the literature

There is disagreement among experts over whether or not the 2 °C target can be met. For example, according to Anderson and Bows (2011), "there is little to no chance" of meeting the target. On the other hand, according to Alcamo *et al.* (2013):

- Policies adopted by parties to the UNFCCC are too weak to meet a 2 or 1.5 °C target. However, these targets might still be achievable if more stringent mitigation policies are adopted immediately.

- Cost-effective 2 °C scenarios project annual global greenhouse gas emissions to peak before the year 2020, with deep cuts in emissions thereafter, leading to a reduction in 2050 of 41% compared to 1990 levels.

Discussion on other targets

Scientific analysis can provide information on the impacts of climate change and associated policies, such as reducing GHG emissions. However, deciding what policies are best requires value judgements. For example, limiting global warming to 1 °C relative to pre-industrial levels may help to reduce climate change damages more than a 2 °C limit. However, a 1 °C limit may be more costly to achieve than a 2 °C limit.

According to some analysts, the 2 °C "guardrail" is inadequate for the needed degree and timeliness of mitigation. On the other hand, some economic studies suggest more modest mitigation policies. For example, the emissions reductions proposed by Nordhaus (2010) might lead to global warming (in the year 2100) of around 3 °C, relative to pre-industrial levels.

Official long-term target of 1.5 °C

In 2015, two official UNFCCC scientific expert bodies came to the conclusion that, "in some regions and vulnerable ecosystems, high risks are projected even for warming above 1.5°C". This expert position was, together with the strong diplomatic voice of the poorest countries and the island nations in the Pacific, the driving force leading to the decision of the Paris Conference 2015, to lay down this 1.5 °C long-term target on top of the existing 2 °C goal.

Encouraging Use Changes

Emissions Tax

An emissions tax on greenhouse gas emissions requires individual emitters to pay a fee, charge or tax for every tonne of greenhouse gas released into the atmosphere. Most environmentally related taxes with implications for greenhouse gas emissions in OECD countries are levied on energy products and motor vehicles, rather than on CO_2 emissions directly.

Emission taxes can be both cost-effective and environmentally effective. Difficulties with emission taxes include their potential unpopularity, and the fact that they cannot guarantee a particular level of emissions reduction. Emissions or energy taxes also often fall disproportionately on lower income classes. In developing countries, institutions may be insufficiently developed for the collection of emissions fees from a wide variety of sources.

Subsidies

According to Mark Z. Jacobson, a program of subsidization balanced against expected flood costs could pay for conversion to 100% renewable power by 2030. Jacobson, and his colleague Mark Delucchi, suggest that the cost to generate and transmit power in 2020 will be less than 4 cents per kilowatt hour (in 2007 dollars) for wind, about 4 cents for wave and hydroelectric, from 4 to 7 cents for geothermal, and 8 cents per kWh for solar, fossil, and nuclear power.

Investment

Another indirect method of encouraging uses of renewable energy, and pursue sustainability and environmental protection, is that of prompting investment in this area through legal means, something that is already being done at national level as well as in the field of international investment.

Carbon Emissions Trading

With the creation of a market for trading carbon dioxide emissions within the Kyoto Protocol, it is likely that London financial markets will be the centre for this potentially highly lucrative business; the New York and Chicago stock markets may have a lower trade volume than expected as long as the US maintains its rejection of the Kyoto.

However, emissions trading may delay the phase-out of fossil fuels.

In the north-east United States, a successful cap and trade program has shown potential for this solution.

The European Union Emission Trading Scheme (EU ETS) is the largest multi-national, greenhouse gas emissions trading scheme in the world. It commenced operation on 1 January 2005, and all 28 member states of the European Union participate in the scheme which has created a new market in carbon dioxide allowances estimated at 35 billion Euros (US$43 billion) per year. The Chicago Climate Exchange was the first (voluntary) emissions market, and is soon to be followed by Asia's first market (Asia Carbon Exchange). A total of 107 million metric tonnes of carbon dioxide equivalent have been exchanged through projects in 2004, a 38% increase relative to 2003 (78 Mt CO_2e).

Twenty three multinational corporations have come together in the G8 Climate Change Roundtable, a business group formed at the January 2005 World Economic Forum. The group includes Ford, Toyota, British Airways and BP. On 9 June 2005 the Group published a statement stating that there was a need to act on climate change and claiming that market-based solutions can help. It called on governments to establish "clear, transparent, and consistent price signals" through "creation of a long-term policy framework" that would include all major producers of greenhouse gases.

The Regional Greenhouse Gas Initiative is a proposed carbon trading scheme being created by nine North-eastern and Mid-Atlantic American states; Connecticut, Delaware, Maine, Massachusetts, New Hampshire, New Jersey, New York, Rhode Island and Vermont. The scheme was due to be developed by April 2005 but has not yet been completed.

Implementation

Implementation puts into effect climate change mitigation strategies and targets. These can be targets set by international bodies or voluntary action by individuals or institutions. This is the most important, expensive and least appealing aspect of environmental governance.

Funding

Implementation requires funding sources but is often beset by disputes over who should provide funds and under what conditions. A lack of funding can be a barrier to successful strategies as there are no formal arrangements to finance climate change development and implementation. Funding is often provided by nations, groups of nations and increasingly NGO and private sources. These funds are often channelled through the Global Environmental Facility (GEF). This is an environmental funding mechanism in the World Bank which is designed to deal with global environmental issues. The GEF was originally designed to tackle four main areas: biological diversity, climate change, international waters and ozone layer depletion, to which land degradation and persistent organic pollutant were added. The GEF funds projects that are agreed to achieve global environmental benefits that are endorsed by governments and screened by one of the GEF's implementing agencies.

Problems

There are numerous issues which result in a current perceived lack of implementation. It has been suggested that the main barriers to implementation are, Uncertainty, Fragmentation, Institutional void, Short time horizon of policies and politicians and Missing motives and willingness to start adapting. The relationships between many climatic processes can cause large levels of uncertainty as they are not fully understood and can be a barrier to implementation. When information on climate change is held between the large numbers of actors involved it can be highly dispersed, context specific or difficult to access causing fragmentation to be a barrier. Institutional void is the lack of commonly accepted rules and norms for policy processes to take place, calling into question the legitimacy and efficacy of policy processes. The Short time horizon of policies and politicians often means that climate change policies are not implemented in favour of socially favoured societal issues. Statements are often posed to keep the illusion of political action to prevent or postpone decisions being made. Missing motives and willingness to start adapting is a large barrier as it prevents any implementation.

The issues that arise with a system which involves international government cooperation, such as Cap and Trade, could potentially be improved with a polycentric approach where the rules are enforced by many small sections of authority as apposed to one overall enforcement agency.

Occurrence

Despite a perceived lack of occurrence, evidence of implementation is emerging internationally. Some examples of this are the initiation of NAPA's and of joint implementation. Many developing nations have made National Adaptation Programs of Action (NAPAs) which are frameworks to prioritize adaption needs. The implementation of many of these is supported by GEF agencies. Many developed countries are implementing 'first generation' institutional adaption plans particularly at the state and local government scale. There has also been a push towards joint implementation between countries by the UNFCC as this has been suggested as a cost-effective way for objectives to be achieved.

United States

Efforts to reduce greenhouse gas emissions by the United States include energy policies which encourage efficiency through programs like Energy Star, Commercial Building Integration, and the Industrial Technologies Program. On 12 November 1998, Vice President Al Gore symbolically signed the Kyoto Protocol, but he indicated participation by the developing nations was necessary prior its being submitted for ratification by the United States Senate.

In 2007, Transportation Secretary Mary Peters, with White House approval, urged governors and dozens of members of the House of Representatives to block California's first-in-the-nation limits on greenhouse gases from cars and trucks, according to e-mails obtained by Congress. The US Climate Change Science Program is a group of about twenty federal agencies and US Cabinet Departments, all working together to address global warming.

The Bush administration pressured American scientists to suppress discussion of global warming, according to the testimony of the Union of Concerned Scientists to the Oversight and Government Reform Committee of the US House of Representatives. "High-quality science" was "struggling to get out," as the Bush administration pressured scientists to tailor their writings on global warming to fit the Bush administration's skepticism, in some cases at the behest of an ex-oil industry lobbyist. "Nearly half of all respondents perceived or personally experienced pressure to eliminate the words 'climate change,' 'global warming' or other similar terms from a variety of communications." Similarly, according to the testimony of senior officers of the Government Accountability Project, the White House attempted to bury the report "National Assessment of the Potential Consequences of Climate Variability and Change," produced by US scientists pursuant to US law. Some US scientists resigned their jobs rather than give in to White House pressure to underreport global warming.

In the absence of substantial federal action, state governments have adopted emissions-control laws such as the Regional Greenhouse Gas Initiative in the Northeast and the Global Warming Solutions Act of 2006 in California.

Developing Countries

In order to reconcile economic development with mitigating carbon emissions, developing countries need particular support, both financial and technical. One of the means of achieving this is the Kyoto Protocol's Clean Development Mechanism (CDM). The World Bank's Prototype Carbon Fund is a public private partnership that operates within the CDM.

An important point of contention, however, is how overseas development assistance not directly related to climate change mitigation is affected by funds provided to climate change mitigation. One of the outcomes of the UNFCC Copenhagen Climate Conference was the Copenhagen Accord, in which developed countries promised to provide US$30 million between 2010 and 2012 of new and additional resources. Yet it remains unclear what exactly the definition of additional is and the European Commission has requested its member states to define what they understand to be additional, and researchers at the Overseas Development Institute have found four main understandings:

1. Climate finance classified as aid, but additional to (over and above) the '0.7%' ODA target;

2. Increase on previous year's Official Development Assistance (ODA) spent on climate change mitigation;

3. Rising ODA levels that include climate change finance but where it is limited to a specified percentage; and

4. Increase in climate finance not connected to ODA.

The main point being that there is a conflict between the OECD states budget deficit cuts, the need to help developing countries adapt to develop sustainably and the need to ensure that funding does not come from cutting aid to other important Millennium Development Goals.

However, none of these initiatives suggest a quantitative cap on the emissions from developing countries. This is considered as a particularly difficult policy proposal as the economic growth of developing countries are proportionally reflected in the growth of greenhouse emissions. Critics of mitigation often argue that, the developing countries' drive to attain a comparable living standard to the developed countries would doom the attempt at mitigation of global warming. Critics also argue that holding down emissions would shift the human cost of global warming from a general one to one that was borne most heavily by the poorest populations on the planet.

In an attempt to provide more opportunities for developing countries to adapt clean technologies, UNEP and WTO urged the international community to reduce trade barriers and to conclude the Doha trade round "which includes opening trade in environmental goods and services".

Non-governmental Approaches

While many of the proposed methods of mitigating global warming require governmental funding, legislation and regulatory action, individuals and businesses can also play a part in the mitigation effort.

Choices in Personal Actions and Business Operations

Environmental groups encourage individual action against global warming, often aimed at the consumer. Common recommendations include lowering home heating and cooling usage, burning less gasoline, supporting renewable energy sources, buying local products to reduce transportation, turning off unused devices, and various others.

A geophysicist at Utrecht University has urged similar institutions to hold the vanguard in voluntary mitigation, suggesting the use of communications technologies such as videoconferencing to reduce their dependence on long-haul flights.

Air Travel and Shipment

In 2008, climate scientist Kevin Anderson raised concern about the growing effect of rapidly increasing global air transport on the climate in a paper, and a presentation, suggesting that reversing this trend is necessary to reduce emissions.

Part of the difficulty is that when aviation emissions are made at high altitude, the climate impacts are much greater than otherwise. Others have been raising the related concerns of the increasing hypermobility of individuals, whether traveling for business or pleasure, involving frequent and often long distance air travel, as well as air shipment of goods.

Business Opportunities and Risks

On 9 May 2005 Jeff Immelt, the chief executive of General Electric (GE), announced plans to reduce GE's global warming related emissions by one percent by 2012. "GE said that given its projected growth, those emissions would have risen by 40 percent without such action."

On 21 June 2005 a group of leading airlines, airports and aerospace manufacturers pledged to work together to reduce the negative environmental impact of aviation, including limiting the impact of air travel on climate change by improving fuel efficiency and reducing carbon dioxide emissions of new aircraft by fifty percent per seat kilometre by 2020 from 2000 levels. The group aims to develop a common reporting system for carbon dioxide emissions per aircraft by the end of 2005, and pressed for the early inclusion of aviation in the European Union's carbon emission trading scheme.

Investor Response

Climate change is also a concern for large institutional investors who have a long term time horizon and potentially large exposure to the negative impacts of global warming because of the large geographic footprint of their multi-national holdings. SRI (Socially responsible investing) Funds allow investors to invest in funds that meet high ESG (environmental, social, governance) standards as such funds invest in companies that are aligned with these goals. Proxy firms can be used to draft guidelines for investment managers that take these concerns into account.

Legal Action

In some countries, those affected by climate change may be able to sue major producers. Attempts at litigation have been initiated by entire peoples such as Palau and the Inuit, as well as non-governmental organizations such as the Sierra Club. Although proving that particular weather events are due specifically to global warming may never be possible, methodologies have been developed to show the increased risk of such events caused by global warming.

For a legal action for negligence (or similar) to succeed, "Plaintiffs must show that, more probably than not, their individual injuries were caused by the risk factor in question, as opposed to any other cause. This has sometimes been translated to a requirement of a relative risk of at least two." Another route (though with little legal bite) is the World Heritage Convention, if it can be shown that climate change is affecting World Heritage Sites like Mount Everest.

Besides countries suing one another, there are also cases where people in a country have taken legal steps against their own government. Legal action for instance has been taken to try to force the US Environmental Protection Agency to regulate greenhouse gas emissions under the Clean Air Act, and against the Export-Import Bank and OPIC for failing to assess environmental impacts (including global warming impacts) under NEPA.

In the Netherlands and Belgium, organisations as Urgenda and the vzw Klimaatzaak in Belgium have also sued their governments as they believe their governments aren't meeting the emission reductions they agreed to. Urgenda has all ready won their case against the Dutch government.

According to a 2004 study commissioned by Friends of the Earth, ExxonMobil and its predecessors caused 4.7 to 5.3 percent of the world's man-made carbon dioxide emissions between 1882 and 2002. The group suggested that such studies could form the basis for eventual legal action.

In 2015, Exxon, received a subpoena. According to the Washington Post and confirmed by the company, the attorney general of New York, Eric Schneiderman, opened an investigation into the possibility that the company had mislead the public and investors about the risks of climate change.

References

- Dan Leon."The Jewish National Fund: How the Land Was 'Redeemed': The JNF's historical concept of exclusively Jewish land is wholly anachronistic"; Palestine-Israel Journal, Vol 12 No. 4 & Vol 13 No. 1, 05/06

- Ceballos, G.; Ehrlich, A. H.; Ehrlich, P. R. (2015). The Annihilation of Nature: Human Extinction of Birds and Mammals. Baltimore, Maryland: Johns Hopkins University Press. pp. 146 ISBN 1421417189

- "SAFnet Dictionary | Definition For [afforestation]". Dictionaryofforestry.org. 2008-10-23. Archived from the original on 2012-03-14. Retrieved 2012-02-17

- Miguel Martínez-Ramos, Elena Alvarez-Buylla and José Sarukhán (June 1989). "Tree Demography and Gap Dynamics in a Tropical Rain Forest". Ecology. 70 (3): 555–558. doi:10.2307/1940203. CS1 maint: Uses authors parameter (link)

- Diesendorf, Mark (2009). Climate action: a campaign manual for greenhouse solutions. Sydney: University of New South Wales Press. p. 116. ISBN 978-1-74223-018-4

- Forest Practices Board. 2007. Lodgepole Pine Stand Structure 25 Years after Mountain Pine Beetle Attack. "Archived copy" (PDF). Archived from the original (PDF) on 2008-10-29. Retrieved 2007-05-12

- Buma, B.; Wessman, C. A. (2011). "Disturbance interactions can impact resilience mechanisms of forests". Ecosphere. 2 (5): art64. doi:10.1890/ES11-00038.1

- Taylor A.F, Kuo F.E, Sullivan W.C (2001). Views of Nature and Self Discipline: Evidence from Inner City Children in Journal of Environmental Psychology (2001), vol. 21

- Adam, David (2009-02-18). "Fifth of world carbon emissions soaked up by extra forest growth, scientists find". The Guardian. London. Retrieved 2010-05-22

- Sousa, W (1984). "The Role of Disturbance in Natural Communities". Annual Review of Ecology and Systematics. 15: 353–391. doi:10.1146/annurev.es.15.110184.002033

- McPherson, G., Simpson, J. R., Peper, P. J., Maco, S. E. & Xiao, Q. 2005. Municipal Forest Benefits and Costs in Five US Cities. Journal of Forestry

- Andrews, Peter (2008). Beyond the brink: Peter Andrews' radical vision for a sustainable Australian landscape. Sydney: ABC Books for the Australian Broadcasting Corporation. p. 40. ISBN 0-7333-2410-X

- K.T. Weber, B.S. Gokhale, (2011). "Effect of grazing on soil-water content in semiarid rangelands of southeast Idaho" Journal of Arid Environments. 75, 464-470

- Savory, Allan; Jody Butterfield (1998-12-01) [1988]. Holistic Management: A New Framework for Decision Making (2nd ed. ed.). Washington, D.C.: Island Press. ISBN 1-55963-487-1

- "Stormwater to Street Trees" (PDF). United States Environmental Protection Agency. United States Environmental Protection Agency. Retrieved September 4, 2015

- Canadell, J.G.; M.R. Raupach (2008-06-13). "Managing Forests for Climate Change". Science. AAAS. 320 (5882): 1456–1457. PMID 18556550. doi:10.1126/science.1155458

- Retallack, Gregory (2001). "Cenozoic Expansion of Grasslands and Climatic Cooling" (PDF). The Journal of Geology. University of Chicago Press. 109: 407–426. Bibcode:2001JG....109..407R. doi:10.1086/320791

- Dale B, Kim S (2004). "Cumulative Energy and Global Warming Impact from the Production of Biomass for Biobased Products". Journal of Industrial Ecology. 7 (3-4): 147–62. doi:10.1162/108819803323059442

- UK Royal Society (September 2009), Geoengineering the climate: science, governance and uncertainty (PDF), London: UK Royal Society, ISBN 978-0-85403-773-5 , RS Policy document 10/09. Report website

- "Forest Service Chief testifies before Senate appropriations committee on 2013 agency budget". US Forest Service. 18 April 2012. Retrieved 29 April 2012

- Parr JF, Sullivan LA (2005). "Soil carbon sequestration in phytoliths". Soil Biology and Biochemistry. 37: 117–24. doi:10.1016/j.soilbio.2004.06.013

- Stern, Nicholas Herbert (2007). The economics of climate change: the Stern review. Cambridge, UK: Cambridge University Press. p. xxv. ISBN 0-521-70080-9

- MBD Energy Ltd. MBD captures Loy Yang Carbon Emissions. Eco Investor June 2009 "Archived copy". Archived from the original on 2011-07-14. Retrieved 2010-01-27. accessed 28 Jan 2010

3

Problems of Deforestation

Deforestation is the clearing of forests to convert them into farms, ranches and other non-forest areas. Apart from human interference, the natural causes of deforestation are drought, soil erosion and flash flood. Desertification, habitat destruction, global warming, arctic sea ice decline and future sea level are some of the topics discussed in this section. The aspects elucidated in this chapter are of vital importance, and provide a better understanding of forestry.

Deforestation

Satellite photograph of deforestation in progress in eastern Bolivia. Worldwide, 10 percent of wilderness areas were lost between 1990 and 2015.

Deforestation, clearance or clearing is the removal of a forest or stand of trees where the land is thereafter converted to a non-forest use. Examples of deforestation include conversion of forest-land to farms, ranches, or urban use. The most concentrated deforestation occurs in tropical rainforests. About 30% of Earth's land surface is covered by forests.

Deforestation occurs for multiple reasons: trees are cut down to be used for building or sold as fuel, (sometimes in the form of charcoal or timber), while cleared land is used as pasture for livestock and plantation. The removal of trees without sufficient reforestation has resulted in damage to habitat, biodiversity loss and aridity. It has adverse impacts on biosequestration of atmospheric carbon dioxide. Deforestation has also been used in war to deprive the enemy of vital resources and cover for its forces. Modern examples of this were the use of Agent Orange by the British military in Malaya during the Malayan Emergency and the United States military in Vietnam during the Vietnam War. As of 2005, net deforestation rates have ceased to increase in countries with a per capita GDP of at least US$4,600. Deforested regions typically incur significant adverse soil erosion and frequently degrade into wasteland.

Disregard of ascribed value, lax forest management and deficient environmental laws are some of the factors that allow deforestation to occur on a large scale. In many countries, deforestation, both naturally occurring and human-induced, is an ongoing issue. Deforestation causes extinction, changes to climatic conditions, desertification, and displacement of populations as observed by current conditions and in the past through the fossil record. More than half of all plant and land animal species in the world live in tropical forests.

Between 2000 and 2012, 2.3 million square kilometres (890,000 square miles) of forests around the world were cut down. As a result of deforestation, only 6.2 million square kilometres (2.4 million square miles) remain of the original 16 million square kilometres (6 million square miles) of forest that formerly covered the Earth.

Causes

According to the United Nations Framework Convention on Climate Change (UNFCCC) secretariat, the overwhelming direct cause of deforestation is agriculture. Subsistence farming is responsible for 48% of deforestation; commercial agriculture is responsible for 32%; logging is responsible for 14%, and fuel wood removals make up 5%.

Experts do not agree on whether industrial logging is an important contributor to global deforestation. Some argue that poor people are more likely to clear forest because they have no alternatives, others that the poor lack the ability to pay for the materials and labour needed to clear forest. One study found that population increases due to high fertility rates were a primary driver of tropical deforestation in only 8% of cases.

Other causes of contemporary deforestation may include corruption of government institutions, the inequitable distribution of wealth and power, population growth and overpopulation, and urbanization. Globalization is often viewed as another root cause of deforestation, though there are cases in which the impacts of globalization (new flows of labor, capital, commodities, and ideas) have promoted localized forest recovery.

The last batch of sawnwood from the peat forest in Indragiri Hulu, Sumatra, Indonesia. Deforestation for oil palm plantation.

In 2000 the United Nations Food and Agriculture Organization (FAO) found that "the role of population dynamics in a local setting may vary from decisive to negligible," and that deforestation can result from "a combination of population pressure and stagnating economic, social and technological conditions."

The degradation of forest ecosystems has also been traced to economic incentives that make forest conversion appear more profitable than forest conservation. Many important forest functions have no markets, and hence, no economic value that is readily apparent to the forests' owners or the communities that rely on forests for their well-being. From the perspective of the developing world, the benefits of forest as carbon sinks or biodiversity reserves go primarily to richer developed nations and there is insufficient compensation for these services. Developing countries feel that some countries in the developed world, such as the United States of America, cut down their forests centuries ago and benefited economically from this deforestation, and that it is hypocritical to deny developing countries the same opportunities, i.e. that the poor shouldn't have to bear the cost of preservation when the rich created the problem.

Some commentators have noted a shift in the drivers of deforestation over the past 30 years. Whereas deforestation was primarily driven by subsistence activities and government-sponsored development projects like transmigration in countries like Indonesia and colonization in Latin America, India, Java, and so on, during the late 19th century and the earlier half of the 20th century, by the 1990s the majority of deforestation was caused by industrial factors, including extractive industries, large-scale cattle ranching, and extensive agriculture.

Problems with Deforestation

Atmospheric

Illegal "slash-and-burn" practice in Madagascar, 2010

Deforestation is ongoing and is shaping climate and geography.

Deforestation is a contributor to global warming, and is often cited as one of the major causes of the enhanced greenhouse effect. Tropical deforestation is responsible for approximately 20% of world greenhouse gas emissions. According to the Intergovernmental Panel on Climate Change deforestation, mainly in tropical areas, could account for up to one-third of total anthropogenic carbon dioxide emissions. But recent calculations suggest that carbon dioxide emissions from deforestation and forest degradation (excluding peatland emissions) contribute about 12% of total anthropogenic carbon dioxide emissions with a range from 6 to 17%. Deforestation causes carbon dioxide to linger in the atmosphere. As carbon dioxide accrues, it produces a layer in the atmosphere that traps radiation from the sun. The radiation converts to heat which causes global warming, which is better known as the greenhouse effect. Plants remove carbon in the form of carbon dioxide from the atmosphere during the process of photosynthesis, but release some carbon dioxide back into the atmosphere during normal respiration. Only when actively growing can a tree or forest remove carbon, by storing it in plant tissues. Both the decay and burning of wood releases much of this stored carbon back to the atmosphere. In order for forests to take up carbon, there must be a net accumulation of wood. One way is for the wood to be harvested and turned into long-lived products, with new young trees replacing them. Deforestation may also cause carbon stores held in soil to be released. Forests can be either sinks or sources depending upon environmental circumstances. Mature forests alternate between being net sinks and net sources of carbon dioxide.

In deforested areas, the land heats up faster and reaches a higher temperature, leading to localized upward motions that enhance the formation of clouds and ultimately produce more rainfall. However, according to the Geophysical Fluid Dynamics Laboratory, the models used to investigate remote responses to tropical deforestation showed a broad but mild temperature increase all through the tropical atmosphere. The model predicted <0.2 °C warming for upper air at 700 mb and 500 mb. However, the model shows no significant changes in other areas besides the Tropics. Though the model showed no significant changes to the climate in areas other than the Tropics, this may not be the case since the model has possible errors and the results are never absolutely definite.

Fires on Borneo and Sumatra, 2006. People use slash-and-burn deforestation to clear land for agriculture.

Reducing emissions from deforestation and forest degradation (REDD) in developing countries has emerged as a new potential to complement ongoing climate policies. The idea consists in providing financial compensations for the reduction of greenhouse gas (GHG) emissions from deforestation and forest degradation".

Rainforests are widely believed by laymen to contribute a significant amount of the world's oxygen, although it is now accepted by scientists that rainforests contribute little net oxygen to the atmosphere and deforestation has only a minor effect on atmospheric oxygen levels. However, the incineration and burning of forest plants to clear land releases large amounts of CO_2, which contributes to global warming. Scientists also state that tropical deforestation releases 1.5 billion tons of carbon each year into the atmosphere.

Hydrological

The water cycle is also affected by deforestation. Trees extract groundwater through their roots and release it into the atmosphere. When part of a forest is removed, the trees no longer transpire this water, resulting in a much drier climate. Deforestation reduces the content of water in the soil and groundwater as well as atmospheric moisture. The dry soil leads to lower water intake for the trees to extract. Deforestation reduces soil cohesion, so that erosion, flooding and landslides ensue.

Shrinking forest cover lessens the landscape's capacity to intercept, retain and transpire precipitation. Instead of trapping precipitation, which then percolates to groundwater systems, deforested areas become sources of surface water runoff, which moves much faster than subsurface flows. Forests return most of the water that falls as precipitation to the atmosphere by transpiration. In contrast, when an area is deforested, almost all precipitation is lost as run-off. That quicker transport of surface water can translate into flash flooding and more localized floods than would occur with the forest cover. Deforestation also contributes to decreased evapotranspiration, which lessens atmospheric moisture which in some cases affects precipitation levels downwind from the deforested area, as water is not recycled to downwind forests, but is lost in runoff and returns directly to the oceans. According to one study, in deforested north and northwest China, the average annual precipitation decreased by one third between the 1950s and the 1980s.

Trees, and plants in general, affect the water cycle significantly:

- their canopies intercept a proportion of precipitation, which is then evaporated back to the atmosphere (canopy interception);

- their litter, stems and trunks slow down surface runoff;

- their roots create macropores – large conduits – in the soil that increase infiltration of water;

- they contribute to terrestrial evaporation and reduce soil moisture via transpiration;

- their litter and other organic residue change soil properties that affect the capacity of soil to store water.

- their leaves control the humidity of the atmosphere by transpiring. 99% of the water absorbed by the roots moves up to the leaves and is transpired.

As a result, the presence or absence of trees can change the quantity of water on the surface, in the soil or groundwater, or in the atmosphere. This in turn changes erosion rates and the availability of water for either ecosystem functions or human services.

The forest may have little impact on flooding in the case of large rainfall events, which overwhelm the storage capacity of forest soil if the soils are at or close to saturation.

Tropical rainforests produce about 30% of our planet's fresh water.

Soil

Undisturbed forests have a very low rate of soil loss (erosion), approximately 2 metric tons per square kilometer (6 short tons per square mile). Deforestation generally increases rates of soil loss, by increasing the amount of runoff and reducing the protection of the soil from tree litter. This can be an advantage in excessively leached tropical rain forest soils. Forestry operations themselves also increase erosion through the development of (forest) roads and the use of mechanized equipment.

China's Loess Plateau was cleared of forest millennia ago. Since then it has been eroding, creating dramatic incised valleys, and providing the sediment that gives the Yellow River its yellow color and that causes the flooding of the river in the lower reaches (hence the river's nickname 'China's sorrow').

Deforestation for the use of clay in the Brazilian city of Rio de Janeiro. The hill depicted is Morro da Covanca, in Jacarepaguá

Removal of trees does not always increase erosion rates. In certain regions of southwest US, shrubs and trees have been encroaching on grassland. The trees themselves enhance the loss of grass between tree canopies. The bare intercanopy areas become highly erodible. The US Forest Service, in Bandelier National Monument for example, is studying how to restore the former ecosystem, and reduce erosion, by removing the trees.

Tree roots bind soil together, and if the soil is sufficiently shallow they act to keep the soil in place by also binding with underlying bedrock. Tree removal on steep slopes with shallow soil thus increases the risk of landslides, which can threaten people living nearby.

Biodiversity

Deforestation on a human scale results in decline in biodiversity, and on a natural global scale is known to cause the extinction of many species.{ The removal or destruction of areas of forest cover has resulted in a degraded environment with reduced biodiversity. Forests support biodiversity, providing habitat for wildlife; moreover, forests foster medicinal conservation. With forest biotopes being irreplaceable source of new drugs (such as taxol), deforestation can destroy genetic variations (such as crop resistance) irretrievably.

Illegal logging in Madagascar. In 2009, the vast majority of the illegally obtained rosewood was exported to China.

Since the tropical rainforests are the most diverse ecosystems on Earth and about 80% of the world's known biodiversity could be found in tropical rainforests, removal or destruction of significant areas of forest cover has resulted in a degraded environment with reduced biodiversity. A study in Rondônia, Brazil, has shown that deforestation also removes the microbial community which is involved in the recycling of nutrients, the production of clean water and the removal of pollutants.

It has been estimated that we are losing 137 plant, animal and insect species every single day due to rainforest deforestation, which equates to 50,000 species a year. Others state that tropical rainforest deforestation is contributing to the ongoing Holocene mass extinction. The known extinction rates from deforestation rates are very low, approximately 1 species per year from mammals and birds which extrapolates to approximately 23,000 species per year for all species. Predictions have been made that more than 40% of the animal and plant species in Southeast Asia could be wiped out in the 21st century. Such predictions were called into question by 1995 data that show that within regions of Southeast Asia much of the original forest has been converted to monospecific plantations, but that potentially endangered species are few and tree flora remains widespread and stable.

Scientific understanding of the process of extinction is insufficient to accurately make predictions about the impact of deforestation on biodiversity. Most predictions of forestry related biodiversity loss are based on species-area models, with an underlying assumption that as the forest declines species diversity will decline similarly. However, many such models have been proven to be wrong and loss of habitat does not necessarily lead to large scale loss of species. Species-area models are known to overpredict the number of species known to be threatened in areas where actual deforestation is ongoing, and greatly overpredict the number of threatened species that are widespread.

A recent study of the Brazilian Amazon predicts that despite a lack of extinctions thus far, up to 90 percent of predicted extinctions will finally occur in the next 40 years.

Economic Impact

Damage to forests and other aspects of nature could halve living standards for the world's poor and reduce global GDP by about 7% by 2050, a report concluded at the Convention on Biological Diversity (CBD) meeting in Bonn in 2008. Historically, utilization of forest products, including timber and fuel wood, has played a key role in human societies, comparable to the roles of water and cultivable land. Today, developed countries continue to utilize timber for building houses, and wood pulp for paper. In developing countries almost three billion people rely on wood for heating and cooking.

The forest products industry is a large part of the economy in both developed and developing countries. Short-term economic gains made by conversion of forest to agriculture, or over-exploitation of wood products, typically leads to loss of long-term income and long-term biological productivity. West Africa, Madagascar, Southeast Asia and many other regions have experienced lower revenue because of declining timber harvests. Illegal logging causes billions of dollars of losses to national economies annually.

The new procedures to get amounts of wood are causing more harm to the economy and overpower the amount of money spent by people employed in logging. According to a study, "in most areas studied, the various ventures that prompted deforestation rarely generated more than US$5 for every ton of carbon they released and frequently returned far less than US$1". The price on the European market for an offset tied to a one-ton reduction in carbon is 23 euro (about US$35).

Rapidly growing economies also have an effect on deforestation. Most pressure will come from the world's developing countries, which have the fastest-growing populations and most rapid economic (industrial) growth. In 1995, economic growth in developing countries reached nearly 6%, compared with the 2% growth rate for developed countries." As our human population grows, new homes, communities, and expansions of cities will occur. Connecting all of the new expansions will be roads, a very important part in our daily life. Rural roads promote economic development but also facilitate deforestation. About 90% of the deforestation has occurred within 100 km of roads in most parts of the Amazon.

The European Union is one of the largest importer of products made from illegal deforestation.

Forest Transition Theory

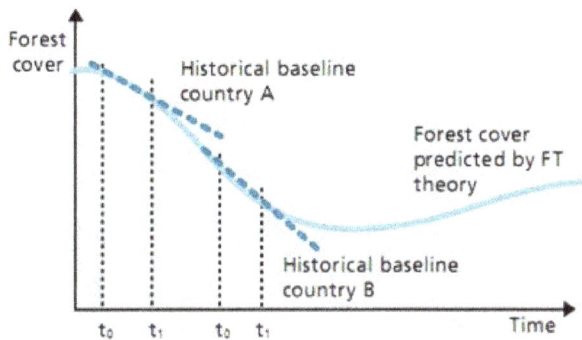

Source: Angelsen 2008.

The forest transition and historical baselines.

The forest area change may follow a pattern suggested by the forest transition (FT) theory, whereby at early stages in its development a country is characterized by high forest cover and low deforestation rates (HFLD countries).

Then deforestation rates accelerate (HFHD, high forest cover – high deforestation rate), and forest cover is reduced (LFHD, low forest cover – high deforestation rate), before the deforestation rate slows (LFLD, low forest cover – low deforestation rate), after which forest cover stabilizes and eventually starts recovering. FT is not a "law of nature," and the pattern is influenced by national context (for example, human population density, stage of development, structure of the economy), global economic forces, and government policies. A country may reach very low levels of forest cover before it stabilizes, or it might through good policies be able to "bridge" the forest transition.

FT depicts a broad trend, and an extrapolation of historical rates therefore tends to underestimate future BAU deforestation for counties at the early stages in the transition (HFLD), while it tends to overestimate BAU deforestation for countries at the later stages (LFHD and LFLD).

Countries with high forest cover can be expected to be at early stages of the FT. GDP per capita captures the stage in a country's economic development, which is linked to the pattern of natural resource use, including forests. The choice of forest cover and GDP per capita also fits well with the two key scenarios in the FT:

(i) a forest scarcity path, where forest scarcity triggers forces (for example, higher prices of forest products) that lead to forest cover stabilization; and

(ii) an economic development path, where new and better off-farm employment opportunities associated with economic growth (= increasing GDP per capita) reduce profitability of frontier agriculture and slows deforestation.

Prehistory

The Carboniferous Rainforest Collapse was an event that occurred 300 million years ago. Climate change devastated tropical rainforests causing the extinction of many plant and animal species. The change was abrupt, specifically, at this time climate became cooler and drier, conditions that are not favourable to the growth of rainforests and much of the biodiversity within them. Rainforests were fragmented forming shrinking 'islands' further and further apart. This sudden collapse affected several large groups, effects on amphibians were particularly devastating, while reptiles fared better, being ecologically adapted to the drier conditions that followed.

An array of Neolithic artifacts, including bracelets, axe heads, chisels, and polishing tools.

Rainforests once covered 14% of the earth's land surface; now they cover a mere 6% and experts estimate that the last remaining rainforests could be consumed in less than 40 years. Small scale deforestation was practiced by some societies for tens of thousands of years before the beginnings of civilization. The first evidence of deforestation appears in the Mesolithic period. It was probably used to convert closed forests into more open ecosystems favourable to game animals. With the advent of agriculture, larger areas began to be deforested, and fire became the prime tool to clear land for crops. In Europe there is little solid evidence before 7000 BC. Mesolithic foragers used fire to create openings for red deer and wild boar. In Great Britain, shade-tolerant species such as oak and ash are replaced in the pollen record by hazels, brambles, grasses and nettles. Removal of the forests led to decreased transpiration, resulting in the formation of upland peat bogs. Widespread decrease in elm pollen across Europe between 8400–8300 BC and 7200–7000 BC, starting in southern Europe and gradually moving north to Great Britain, may represent land clearing by fire at the onset of Neolithic agriculture.

The Neolithic period saw extensive deforestation for farming land. Stone axes were being made from about 3000 BC not just from flint, but from a wide variety of hard rocks from across Britain and North America as well. They include the noted Langdale axe industry in the English Lake District, quarries developed at Penmaenmawr in North Wales and numerous other locations. Roughouts were made locally near the quarries, and some were polished locally to give a fine finish. This step not only increased the mechanical strength of the axe, but also made penetration of wood easier. Flint was still used from sources such as Grimes Graves but from many other mines across Europe.

Evidence of deforestation has been found in Minoan Crete; for example the environs of the Palace of Knossos were severely deforested in the Bronze Age.

Pre-industrial History

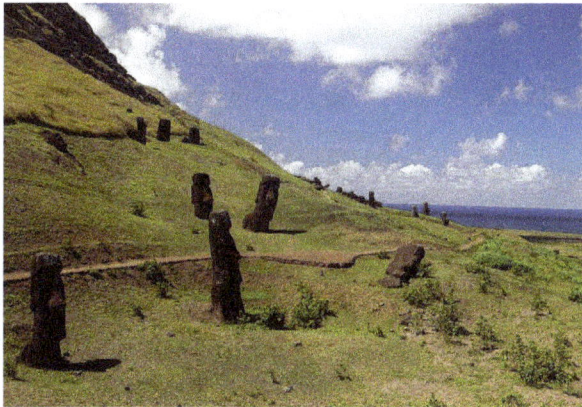

Easter Island, deforested. According to Jared Diamond: "Among past societies faced with the prospect of ruinous deforestation, Easter Island and Mangareva chiefs succumbed to their immediate concerns, but Tokugawa shoguns, Inca emperors, New Guinea highlanders, and 16th century German landowners adopted a long view and reafforested."

Throughout prehistory, humans were hunter gatherers who hunted within forests. In most areas, such as the Amazon, the tropics, Central America, and the Caribbean, only after shortages of wood and other forest products occur are policies implemented to ensure forest resources are used in a sustainable manner.

In ancient Greece, Tjeered van Andel and co-writers summarized three regional studies of historic erosion and alluviation and found that, wherever adequate evidence exists, a major phase of erosion follows, by about 500-1,000 years the introduction of farming in the various regions of Greece, ranging from the later Neolithic to the Early Bronze Age. The thousand years following the mid-first millennium BC saw serious, intermittent pulses of soil erosion in numerous places. The historic silting of ports along the southern coasts of Asia Minor (*e.g.* Clarus, and the examples of Ephesus, Priene and Miletus, where harbors had to be abandoned because of the silt deposited by the Meander) and in coastal Syria during the last centuries BC.

Easter Island has suffered from heavy soil erosion in recent centuries, aggravated by agriculture and deforestation. Jared Diamond gives an extensive look into the collapse of the ancient Easter Islanders in his book *Collapse*. The disappearance of the island's trees seems to coincide with a decline of its civilization around the 17th and 18th century. He attributed the collapse to deforestation and over-exploitation of all resources.

The famous silting up of the harbor for Bruges, which moved port commerce to Antwerp, also followed a period of increased settlement growth (and apparently of deforestation) in the upper river basins. In early medieval Riez in upper Provence, alluvial silt from two small rivers raised the riverbeds and widened the floodplain, which slowly buried the Roman settlement in alluvium and gradually moved new construction to higher ground; concurrently the headwater valleys above Riez were being opened to pasturage.

A typical progress trap was that cities were often built in a forested area, which would provide wood for some industry (for example, construction, shipbuilding, pottery). When deforestation occurs without proper replanting, however; local wood supplies become difficult to obtain near enough to remain competitive, leading to the city's abandonment, as happened repeatedly in Ancient Asia Minor. Because of fuel needs, mining and metallurgy often led to deforestation and city abandonment.

With most of the population remaining active in (or indirectly dependent on) the agricultural sector, the main pressure in most areas remained land clearing for crop and cattle farming. Enough wild green was usually left standing (and partially used, for example, to collect firewood, timber and fruits, or to graze pigs) for wildlife to remain viable. The elite's (nobility and higher clergy) protection of their own hunting privileges and game often protected significant woodlands.

Major parts in the spread (and thus more durable growth) of the population were played by monastical 'pioneering' (especially by the Benedictine and Commercial orders) and some feudal lords' recruiting farmers to settle (and become tax payers) by offering relatively good legal and fiscal conditions. Even when speculators sought to encourage towns, settlers needed an agricultural belt around or sometimes within defensive walls. When populations were quickly decreased by causes such as the Black Death or devastating warfare (for example, Genghis Khan's Mongol hordes in eastern and central Europe, Thirty Years' War in Germany), this could lead to settlements being abandoned. The land was reclaimed by nature, but the secondary forests usually lacked the original biodiversity.

Deforestation of Brazil's Atlantic Forest c.1820–1825

From 1100 to 1500 AD, significant deforestation took place in Western Europe as a result of the expanding human population. The large-scale building of wooden sailing ships by European (coastal) naval owners since the 15th century for exploration, colonisation, slave trade–and other trade on the high seas consumed many forest resources. Piracy also contributed to the over harvesting of forests, as in Spain. This led to a weakening of the domestic economy after Columbus' discovery of America, as the economy became dependent on colonial activities (plundering, mining, cattle, plantations, trade, etc.)

In *Changes in the Land* (1983), William Cronon analyzed and documented 17th-century English colonists' reports of increased seasonal flooding in New England during the period when new settlers initially cleared the forests for agriculture. They believed flooding was linked to widespread forest clearing upstream.

The massive use of charcoal on an industrial scale in Early Modern Europe was a new type of consumption of western forests; even in Stuart England, the relatively primitive production of charcoal has already reached an impressive level. Stuart England was so widely deforested that it depended on the Baltic trade for ship timbers, and looked to the untapped forests of New England to supply the need. Each of Nelson's Royal Navy war ships at Trafalgar (1805) required 6,000 mature oaks for its construction. In France, Colbert planted oak forests to supply the French navy in the future. When the oak plantations matured in the mid-19th century, the masts were no longer required because shipping had changed.

Norman F. Cantor's summary of the effects of late medieval deforestation applies equally well to Early Modern Europe:

Europeans had lived in the midst of vast forests throughout the earlier medieval centuries. After 1250 they became so skilled at deforestation that by 1500 they were running short of wood for heating and cooking. They were faced with a nutritional decline because of the elimination of the generous supply of wild game that had inhabited the now-disappearing forests, which throughout medieval times had provided the staple of their carnivorous high-protein diet. By 1500 Europe was on the edge of a fuel and nutritional disaster [from] which it was saved in the sixteenth century only by the burning of soft coal and the cultivation of potatoes and maize.

Industrial Era

In the 19th century, introduction of steamboats in the United States was the cause of deforestation of banks of major rivers, such as the Mississippi River, with increased and more severe flooding one of the environmental results. The steamboat crews cut wood every day from the riverbanks to fuel the steam engines. Between St. Louis and the confluence with the Ohio River to the south, the Mississippi became more wide and shallow, and changed its channel laterally. Attempts to improve navigation by the use of snag pullers often resulted in crews' clearing large trees 100 to 200 feet (61 m) back from the banks. Several French colonial towns of the Illinois Country, such as Kaskaskia, Cahokia and St. Philippe, Illinois were flooded and abandoned in the late 19th century, with a loss to the cultural record of their archeology.

The wholescale clearance of woodland to create agricultural land can be seen in many parts of the world, such as the Central forest-grasslands transition and other areas of the Great Plains of the United States. Specific parallels are seen in the 20th-century deforestation occurring in many developing nations.

Rates of Deforestation

Slash-and-burn farming in the state of Rondônia, western Brazil

Global deforestation sharply accelerated around 1852. It has been estimated that about half of the Earth's mature tropical forests—between 7.5 million and 8 million km² (2.9 million to 3 million sq mi) of the original 15 million to 16 million km² (5.8 million to 6.2 million sq mi) that until 1947 covered the planet—have now been destroyed. Some scientists have predicted that unless significant measures (such as seeking out and protecting old growth forests that have not been disturbed) are taken on a worldwide basis, by 2030 there will only be 10% remaining, with another 10% in a degraded condition. 80% will have been lost, and with them hundreds of thousands of irreplaceable species. Some cartographers have attempted to illustrate the sheer scale of deforestation by country using a cartogram.

Estimates vary widely as to the extent of tropical deforestation. Scientists estimate that one fifth of the world's tropical rainforest was destroyed between 1960 and 1990. They claim that that rainforests 60 years ago covered 14% of the world's land surface, now only cover 5–7%, and that all tropical forests will be gone by the middle of the 21st century.

A 2002 analysis of satellite imagery suggested that the rate of deforestation in the humid tropics (approximately 5.8 million hectares per year) was roughly 23% lower than the most commonly quoted rates. Conversely, a newer analysis of satellite images reveals that deforestation of the Amazon rainforest is twice as fast as scientists previously estimated.

Some have argued that deforestation trends may follow a Kuznets curve, which if true would nonetheless fail to eliminate the risk of irreversible loss of non-economic forest values (for example, the extinction of species).

A 2005 report by the United Nations Food and Agriculture Organization (FAO) estimated that although the Earth's total forest area continued to decrease at about 13 million hectares per year, the global rate of deforestation has recently been slowing. The 2016 report by the FAO reports from 2010 to 2015 there was a worldwide decrease in forest area of 3.3 million ha per year. During this five-year period, the biggest forest area loss occurred in the tropics, particularly in South America and Africa. Per capita forest area decline was also greatest in the tropics and subtropics but is occurring in every climatic domain (except in the temperate) as populations increase.

Satellite image of Haiti's border with the Dominican Republic (right) shows the amount of deforestation on the Haitian side

Others claim that rainforests are being destroyed at an ever-quickening pace. The London-based Rainforest Foundation notes that "the UN figure is based on a definition of forest as being an area with as little as 10% actual tree cover, which would therefore include areas that are actually savannah-like ecosystems and badly damaged forests." Other critics of the FAO data point out that they do not distinguish between forest types, and that they are based largely on reporting from forestry departments of individual countries, which do not take into account unofficial activities like illegal logging.

Deforestation around Pakke Tiger Reserve, India

Despite these uncertainties, there is agreement that destruction of rainforests remains a significant environmental problem. Up to 90% of West Africa's coastal rainforests have disappeared since 1900. In South Asia, about 88% of the rainforests have been lost. Much of what remains of the world's rainforests is in the Amazon basin, where the Amazon Rainforest covers approximately 4 million square kilometres. The regions with the highest tropical deforestation rate between 2000 and 2005 were Central America—which lost 1.3% of its forests each year—and tropical Asia. In Central America, two-thirds of lowland tropical forests have been turned into pasture since 1950 and 40% of all the rainforests have been lost in the last 40 years. Brazil has lost 90–95% of its Mata Atlântica forest. Paraguay was losing its natural semi humid forests in the country's western regions at a rate of 15.000 hectares at a randomly studied 2-month period in 2010, Paraguay's parliament refused in 2009 to pass a law that would have stopped cutting of natural forests altogether.

Madagascar has lost 90% of its eastern rainforests. As of 2007, less than 50% of Haiti's forests remained. Mexico, India, the Philippines, Indonesia, Thailand, Burma, Malaysia, Bangladesh, China, Sri Lanka, Laos, Nigeria, the Democratic Republic of the Congo, Liberia, Guinea, Ghana and the Ivory Coast, have lost large areas of their rainforest. Several countries, notably Brazil, have declared their deforestation a national emergency. The World Wildlife Fund's ecoregion project catalogues habitat types throughout the world, including habitat loss such as deforestation, showing for example that even in the rich forests of parts of Canada such as the Mid-Continental Canadian forests of the prairie provinces half of the forest cover has been lost or altered.

Regions

Rates of deforestation vary around the world.

In 2011 Conservation International listed the top 10 most endangered forests, characterized by having all lost 90% or more of their original habitat, and each harboring at least 1500 endemic plant species (species found nowhere else in the world).

Top 10 Most Endangered Forests 2011				
Endangered forest	**Region**	**Remaining habitat**	**Predominate vegetation type**	**Notes**
Indo-Burma	Asia-Pacific	5%	Tropical and subtropical moist broadleaf forests	Rivers, floodplain wetlands, mangrove forests. Burma, Thailand, Laos, Vietnam, Cambodia, India.
New Caledonia	Asia-Pacific	5%	Tropical and subtropical moist broadleaf forests	-
Sundaland	Asia-Pacific	7%	Tropical and subtropical moist broadleaf forests	Western half of the Indo-Malayan archipelago including southern Borneo and Sumatra.
Philippines	Asia-Pacific	7%	Tropical and subtropical moist broadleaf forests	Forests over the entire country including 7,100 islands.
Atlantic Forest	South America	8%	Tropical and subtropical moist broadleaf forests	Forests along Brazil's Atlantic coast, extends to parts of Paraguay, Argentina and Uruguay.
Mountains of Southwest China	Asia-Pacific	8%	Temperate coniferous forest	-
California Floristic Province	North America	10%	Tropical and subtropical dry broadleaf forests	-
Coastal Forests of Eastern Africa	Africa	10%	Tropical and subtropical moist broadleaf forests	Mozambique, Tanzania, Kenya, Somalia.

Madagascar & Indian Ocean Islands	Africa	10%	Tropical and subtropical moist broadleaf forests	Madagascar, Mauritius, Reunion, Seychelles, Comoros.
Eastern Afromontane	Africa	11%	Tropical and subtropical moist broadleaf forests Montane grasslands and shrublands	Forests scattered along the eastern edge of Africa, from Saudi Arabia in the north to Zimbabwe in the south.

Control

Reducing Emissions

Main international organizations including the United Nations and the World Bank, have begun to develop programs aimed at curbing deforestation. The blanket term Reducing Emissions from Deforestation and Forest Degradation (REDD) describes these sorts of programs, which use direct monetary or other incentives to encourage developing countries to limit and/or roll back deforestation. Funding has been an issue, but at the UN Framework Convention on Climate Change (UNFCCC) Conference of the Parties-15 (COP-15) in Copenhagen in December 2009, an accord was reached with a collective commitment by developed countries for new and additional resources, including forestry and investments through international institutions, that will approach USD 30 billion for the period 2010–2012. Significant work is underway on tools for use in monitoring developing country adherence to their agreed REDD targets. These tools, which rely on remote forest monitoring using satellite imagery and other data sources, include the Center for Global Development's FORMA (Forest Monitoring for Action) initiative and the Group on Earth Observations' Forest Carbon Tracking Portal. Methodological guidance for forest monitoring was also emphasized at COP-15. The environmental organization Avoided Deforestation Partners leads the campaign for development of REDD through funding from the U.S. government. In 2014, the Food and Agriculture Organization of the United Nations and partners launched Open Foris – a set of open-source software tools that assist countries in gathering, producing and disseminating information on the state of forest resources. The tools support the inventory lifecycle, from needs assessment, design, planning, field data collection and management, estimation analysis, and dissemination. Remote sensing image processing tools are included, as well as tools for international reporting for Reducing emissions from deforestation and forest degradation (REDD) and MRV and FAO's Global Forest Resource Assessments.

In evaluating implications of overall emissions reductions, countries of greatest concern are those categorized as High Forest Cover with High Rates of Deforestation (HFHD) and Low Forest Cover with High Rates of Deforestation (LFHD). Afghanistan, Benin, Botswana, Burma, Burundi, Cameroon, Chad, Ecuador, El Salvador, Ethiopia, Ghana, Guatemala, Guinea, Haiti, Honduras, Indonesia, Liberia, Malawi, Mali, Mauritania, Mongolia, Namibia, Nepal, Nicaragua, Niger, Nigeria, Pakistan, Paraguay, Philippines, Senegal, Sierra Leone, Sri Lanka, Sudan, Togo, Uganda, United Republic of Tanzania, Zimbabwe are listed as having Low Forest Cover with High Rates of Deforestation (LFHD). Brazil, Cambodia, Democratic Peoples Republic of Korea, Equatorial Guinea, Malaysia, Solomon Islands, Timor-Leste, Venezuela, Zambia are listed as High Forest Cover with High Rates of Deforestation (HFHD).

Payments for conserving Forests

In Bolivia, deforestation in upper river basins has caused environmental problems, including soil erosion and declining water quality. An innovative project to try and remedy this situation involves landholders in upstream areas being paid by downstream water users to conserve forests. The landholders receive US$20 to conserve the trees, avoid polluting livestock practices, and enhance the biodiversity and forest carbon on their land. They also receive US$30, which purchases a bee-hive, to compensate for conservation for two hectares of water-sustaining forest for five years. Honey revenue per hectare of forest is US$5 per year, so within five years, the landholder has sold US$50 of honey. The project is being conducted by Fundación Natura Bolivia and Rare Conservation, with support from the Climate & Development Knowledge Network.

Land Rights

Transferring land rights to indigenous inhabitants is argued to efficiently conserve forests.

Transferring rights over land from public domain to its indigenous inhabitants is argued to be a cost effective strategy to conserve forests. This includes the protection of such rights entitled in existing laws, such as India's Forest Rights Act. The transferring of such rights in China, perhaps the largest land reform in modern times, has been argued to have increased forest cover. In Brazil, forested areas given tenure to indigenous groups have even lower rates of clearing than national parks.

Farming

New methods are being developed to farm more intensively, such as high-yield hybrid crops, greenhouse, autonomous building gardens, and hydroponics. These methods are often dependent on chemical inputs to maintain necessary yields. In cyclic agriculture, cattle are grazed on farm land that is resting and rejuvenating. Cyclic agriculture actually increases the fertility of the soil. Intensive farming can also decrease soil nutrients by consuming at an accelerated rate the trace minerals needed for crop growth. The most promising approach, however, is the concept of food forests in permaculture, which consists of agroforestal systems carefully designed to mimic natural forests, with an emphasis on plant and animal species of interest for food, timber and other uses. These systems have low dependence on fossil fuels and agro-chemicals, are highly self-maintaining, highly productive, and with strong positive impact on soil and water quality, and biodiversity.

Monitoring Deforestation

There are multiple methods that are appropriate and reliable for reducing and monitoring deforestation. One method is the "visual interpretation of aerial photos or satellite imagery that is labor-intensive but does not require high-level training in computer image processing or extensive computational resources". Another method includes hot-spot analysis (that is, locations of rapid change) using expert opinion or coarse resolution satellite data to identify locations for detailed digital analysis with high resolution satellite images. Deforestation is typically assessed by quantifying the amount of area deforested, measured at the present time. From an environmental point of view, quantifying the damage and its possible consequences is a more important task, while conservation efforts are more focused on forested land protection and development of land-use alternatives to avoid continued deforestation. Deforestation rate and total area deforested, have been widely used for monitoring deforestation in many regions, including the Brazilian Amazon deforestation monitoring by INPE. A global satellite view is available.

Forest Management

Efforts to stop or slow deforestation have been attempted for many centuries because it has long been known that deforestation can cause environmental damage sufficient in some cases to cause societies to collapse. In Tonga, paramount rulers developed policies designed to prevent conflicts between short-term gains from converting forest to farmland and long-term problems forest loss would cause, while during the 17th and 18th centuries in Tokugawa, Japan, the shoguns developed a highly sophisticated system of long-term planning to stop and even reverse deforestation of the preceding centuries through substituting timber by other products and more efficient use of land that had been farmed for many centuries. In 16th-century Germany, landowners also developed silviculture to deal with the problem of deforestation. However, these policies tend to be limited to environments with *good rainfall*, *no dry season* and *very young soils* (through volcanism or glaciation). This is because on older and less fertile soils trees grow too slowly for silviculture to be economic, whilst in areas with a strong dry season there is always a risk of forest fires destroying a tree crop before it matures.

In the areas where "slash-and-burn" is practiced, switching to "slash-and-char" would prevent the rapid deforestation and subsequent degradation of soils. The biochar thus created, given back to the soil, is not only a durable carbon sequestration method, but it also is an extremely beneficial amendment to the soil. Mixed with biomass it brings the creation of terra preta, one of the richest soils on the planet and the only one known to regenerate itself.

Sustainable Practices

Certification, as provided by global certification systems such as Programme for the Endorsement of Forest Certification and Forest Stewardship Council, contributes to tackling deforestation by creating market demand for timber from sustainably managed forests. According to the United Nations Food and Agriculture Organization (FAO), "A major condition for the adoption of sustainable forest management is a demand for products that are produced sustainably and consumer willingness to pay for the higher costs entailed. Certification represents a shift from regulatory approaches to market incentives to promote sustainable forest management. By promoting the positive attributes of forest products from sustainably managed forests, certification focuses on the demand side of environmental conservation." Rainforest Rescue argues that the standards of organizations like FSC are too closely connected to timber industry interests and therefore do not guarantee environmentally and socially responsible forest management. In reality, monitoring systems are inadequate and various cases of fraud have been documented worldwide.

Bamboo is advocated as a more sustainable alternative for cutting down wood for fuel.

Some nations have taken steps to help increase the amount of trees on Earth. In 1981, China created National Tree Planting Day Forest and forest coverage had now reached 16.55% of China's land mass, as against only 12% two decades ago.

Using fuel from bamboo rather than wood results in cleaner burning, and since bamboo matures much faster than wood, deforestation is reduced as supply can be replenished faster.

Reforestation

In many parts of the world, especially in East Asian countries, reforestation and afforestation are increasing the area of forested lands. The amount of woodland has increased in 22 of the world's 50 most forested nations. Asia as a whole gained 1 million hectares of forest between 2000 and 2005. Tropical forest in El Salvador expanded more than 20% between 1992 and 2001. Based on these trends, one study projects that global forest will increase by 10%—an area the size of India—by 2050.

In the People's Republic of China, where large scale destruction of forests has occurred, the government has in the past required that every able-bodied citizen between the ages of 11 and 60 plant three to five trees per year or do the equivalent amount of work in other forest services. The government claims that at least 1 billion trees have been planted in China every year since 1982. This is no longer required today, but 12 March of every year in China is the Planting Holiday. Also, it has introduced the Green Wall of China project, which aims to halt the expansion of the Gobi desert through the planting of trees. However, due to the large percentage of trees dying off after planting (up to 75%), the project is not very successful. There has been a 47-million-hectare increase in forest area in China since the 1970s. The total number of trees amounted to be about 35 billion and 4.55% of China's land mass increased in forest coverage. The forest coverage was 12% two decades ago and now is 16.55%.

An ambitious proposal for China is the Aerially Delivered Re-forestation and Erosion Control System and the proposed Sahara Forest Project coupled with the Seawater Greenhouse.

In Western countries, increasing consumer demand for wood products that have been produced and harvested in a sustainable manner is causing forest landowners and forest industries to become increasingly accountable for their forest management and timber harvesting practices.

The Arbor Day Foundation's Rain Forest Rescue program is a charity that helps to prevent deforestation. The charity uses donated money to buy up and preserve rainforest land before the lumber companies can buy it. The Arbor Day Foundation then protects the land from deforestation. This also locks in the way of life of the primitive tribes living on the forest land. Organizations such as Community Forestry International, Cool Earth, The Nature Conservancy, World Wide Fund for Nature, Conservation International, African Conservation Foundation and Greenpeace also focus on preserving forest habitats. Greenpeace in particular has also mapped out the forests that are still intact and published this information on the internet. World Resources Institute in turn has made a simpler thematic map showing the amount of forests present just before the age of man (8000 years ago) and the current (reduced) levels of forest. These maps mark the amount of afforestation required to repair the damage caused by people.

Forest Plantations

To meet the world's demand for wood, it has been suggested by forestry writers Botkins and Sedjo that high-yielding forest plantations are suitable. It has been calculated that plantations yielding 10 cubic meters per hectare annually could supply all the timber required for international trade on 5% of the world's existing forestland. By contrast, natural forests produce about 1–2 cubic meters per hectare; therefore, 5–10 times more forestland would be required to meet demand. Forester Chad Oliver has suggested a forest mosaic with high-yield forest lands interspersed with conservation land.

Globally, planted forests increased from 4.1% to 7.0% of the total forest area between 1990 and 2015. Plantation forests made up 280 million ha in 2015, an increase of about 40 million ha in the last ten years. Globally, planted forests consist of about 18% exotic or introduced species while the rest are species native to the country where they are planted. In South America, Oceania, and East and Southern Africa, planted forests are dominated by introduced species: 88%, 75% and 65%, respectively. In North America, West and Central Asia, and Europe the proportions of introduced species in plantations are much lower at 1%, 3% and 8% of the total area planted, respectively.

In the country of Senegal, on the western coast of Africa, a movement headed by youths has helped to plant over 6 million mangrove trees. The trees will protect local villages from storm damages and will provide a habitat for local wildlife. The project started in 2008, and already the Senegalese government has been asked to establish rules and regulations that would protect the new mangrove forests.

Military Context

While the preponderance of deforestation is due to demands for agricultural and urban use for the human population, there are some examples of military causes. One example of deliberate deforestation is that which took place in the U.S. zone of occupation in Germany after World War II. Before the onset of the Cold War, defeated Germany was still considered a potential future threat rather than potential future ally. To address this threat, attempts were made to lower German industrial potential, of which forests were deemed an element. Sources in the U.S. government admitted that the purpose of this was that the "ultimate destruction of the war potential of German forests." As a consequence of the practice of clear-felling, deforestation resulted which could "be replaced only by long forestry development over perhaps a century."

American Sherman tanks knocked out by Japanese artillery on Okinawa.

Deforestation can also be one consequence of war. For example, in the 1945 Battle of Okinawa, bombardment and other combat operations reduced the lush tropical landscape into "a vast field of mud, lead, decay and maggots". Deforestation can also be an intentional tactic of military forces. Defoliants (like Agent Orange or others) was used by the British in the Malayan Emergency, and by the United States in the Korean War and Vietnam War.

Public Health Context

Deforestation eliminates a great number of species of plants and animals which also often results in an increase in disease. Loss of native species allows new species to come to dominance. Often the destruction of predatory species can result in an increase in rodent populations which can carry plague. Additionally, erosion can produce pools of stagnant water that are perfect breeding grounds for mosquitos, well known vectors of malaria, yellow fever, nipah virus, and more. Deforestation can also create a path for non-native species to flourish such as certain types of snails, which have been correlated with an increase in schistosomiasis cases.

Deforestation is occurring all over the world and has been coupled with an increase in the occurrence of disease outbreaks. In Malaysia, thousands of acres of forest have been cleared for pig farms. This has resulted in an increase in the zoonosis the Nipah virus. In Kenya, deforestation has led to an increase in malaria cases which is now the leading cause of morbidity and mortality the country. A 2017 study in the *American Economic Review* found that deforestation substantially increased the incidence of malaria in Nigeria.

Another pathway through which deforestation affects disease is the relocation and dispersion of disease-carrying hosts. This disease emergence pathway can be called "range expansion," whereby the host's range (and thereby the range of pathogens) expands to new geographic areas. Through deforestation, hosts and reservoir species are forced into neighboring habitats. Accompanying the reservoir species are pathogens that have the ability to find new hosts in previously unexposed regions. As these pathogens and species come into closer contact with humans, they are infected both directly and indirectly.

A catastrophic example of range expansion is the 1998 outbreak of Nipah virus in Malaysia. For a number of years, deforestation, drought, and subsequent fires led to a dramatic geographic shift and density of fruit bats, a reservoir for Nipah virus. Deforestation reduced the available fruiting trees in the bats' habitat, and they encroached on surrounding orchards which also happened to be the location of a large number of pigsties. The bats, through proximity spread the Nipah to pigs. While the virus infected the pigs, mortality was much lower than among humans, making the pigs a virulent host leading to the transmission of the virus to humans. This resulted in 265 reported cases of encephalitis, of which 105 resulted in death. This example provides an important lesson for the impact deforestation can have on human health.

Another example of range expansion due to deforestation and other anthropogenic habitat impacts includes the Capybara rodent in Paraguay. This rodent is the host of a number of zoonotic diseases and, while there has not yet been a human-borne outbreak due to the movement of this rodent into new regions, it offers an example of how habitat destruction through deforestation and subsequent movements of species is occurring regularly.

A now well-developed theory is that the spread of HIV it is at least partially due deforestation. Rising populations created a food demand and with deforestation opening up new areas of the forest the hunters harvested a great deal of primate bushmeat, which is believed to be the origin of HIV.

Soil erosion

An actively eroding rill on an intensively-farmed field in eastern Germany

Soil erosion is the displacement of upper layer of soil, one form of soil degradation. The erosion of soil is a naturally occurring process on all land. The agents of soil erosion are water and wind, each contributing a significant amount of soil loss each year. Soil erosion may be a slow process that continues relatively unnoticed, or it may occur at an alarming rate causing a serious loss of topsoil. The loss of soil from farmland may be reflected in reduced crop production potential, lower surface water quality and damaged drainage networks.

While erosion is a natural process, human activities have increased by 10–40 times the rate at which erosion is occurring globally. Excessive (or accelerated) erosion causes both "on-site" and "off-site" problems. On-site impacts include decreases in agricultural productivity and (on natural landscapes) ecological collapse, both because of loss of the nutrient-rich upper soil layers. In some cases, the eventual end result is desertification. Off-site effects include sedimentation of waterways and eutrophication of water bodies, as well as sediment-related damage to roads and houses. Water and wind erosion are the two primary causes of land degradation; combined, they are responsible for about 84% of the global extent of degraded land, making excessive erosion one of the most significant environmental problems worldwide.

Intensive agriculture, deforestation, roads, anthropogenic climate change and urban sprawl are amongst the most significant human activities in regard to their effect on stimulating erosion. However, there are many prevention and remediation practices that can curtail or limit erosion of vulnerable soils.

Physical Processes

Rainfall and Surface Runoff

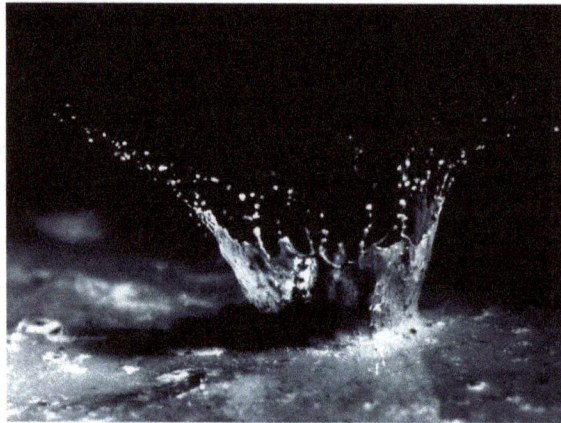

Soil and water being splashed by the impact of a single raindrop.

Rainfall, and the surface runoff which may result from rainfall, produces four main types of soil erosion: *splash erosion*, *sheet erosion*, *rill erosion*, and *gully erosion*. Splash erosion is generally seen as the first and least severe stage in the soil erosion process, which is followed by sheet erosion, then rill erosion and finally gully erosion (the most severe of the four).

In *splash erosion*, the impact of a falling raindrop creates a small crater in the soil, ejecting soil particles. The distance these soil particles travel can be as much as 0.6 m (two feet) vertically and 1.5 m (five feet) horizontally on level ground.

If the soil is saturated, or if the rainfall rate is greater than the rate at which water can infiltrate into the soil, surface runoff occurs. If the runoff has sufficient flow energy, it will transport loosened soil particles (sediment) down the slope. *Sheet erosion* is the transport of loosened soil particles by overland flow.

A spoil tip covered in rills and gullies due to erosion processes caused by rainfall: Rummu, Estonia

Rill erosion refers to the development of small, ephemeral concentrated flow paths which function as both sediment source and sediment delivery systems for erosion on hillslopes. Generally, where water erosion rates on disturbed upland areas are greatest, rills are active. Flow depths in rills are typically of the order of a few centimeters (about an inch) or less and along-channel slopes may be quite steep. This means that rills exhibit hydraulic physics very different from water flowing through the deeper, wider channels of streams and rivers.

Gully erosion occurs when runoff water accumulates and rapidly flows in narrow channels during or immediately after heavy rains or melting snow, removing soil to a considerable depth.

Rivers and Streams

Dobbingstone Burn, Scotland—This photo illustrates two different types of erosion affecting the same place. Valley erosion is occurring due to the flow of the stream, and the boulders and stones (and much of the soil) that are lying on the edges are glacial till that was left behind as ice age glaciers flowed over the terrain.

Valley or *stream erosion* occurs with continued water flow along a linear feature. The erosion is both downward, deepening the valley, and headward, extending the valley into the hillside, creating head cuts and steep banks. In the earliest stage of stream erosion, the erosive activity is dominantly vertical, the valleys have a typical V cross-section and the stream gradient is relatively steep. When some base level is reached, the erosive activity switches to lateral erosion, which widens the valley floor and creates a narrow floodplain. The stream gradient becomes nearly flat, and lateral deposition of sediments becomes important as the stream meanders across the valley floor. In all stages of stream erosion, by far the most erosion occurs during times of flood, when more and faster-moving water is available to carry a larger sediment load. In such processes, it is not the water alone that erodes: suspended abrasive particles, pebbles and boulders can also act erosively as they traverse a surface, in a process known as *traction*.

Bank erosion is the wearing away of the banks of a stream or river. This is distinguished from changes on the bed of the watercourse, which is referred to as *scour*. Erosion and changes in the form of river banks may be measured by inserting metal rods into the bank and marking the position of the bank surface along the rods at different times.

Thermal erosion is the result of melting and weakening permafrost due to moving water. It can occur both along rivers and at the coast. Rapid river channel migration observed in the Lena River of Siberia is due to thermal erosion, as these portions of the banks are composed of permafrost-cemented non-cohesive materials. Much of this erosion occurs as the weakened banks fail in large slumps. Thermal erosion also affects the Arctic coast, where wave action and near-shore temperatures combine to undercut permafrost bluffs along the shoreline and cause them to fail. Annual erosion rates along a 100-kilometre (62-mile) segment of the Beaufort Sea shoreline averaged 5.6 metres (18 feet) per year from 1955 to 2002.

Floods

At extremely high flows, kolks, or vortices are formed by large volumes of rapidly rushing water. Kolks cause extreme local erosion, plucking bedrock and creating pothole-type geographical features called Rock-cut basins. Examples can be seen in the flood regions result from glacial Lake Missoula, which created the channeled scablands in the Columbia Basin region of eastern Washington.

Wind Erosion

Árbol de Piedra, a rock formation in the Altiplano, Bolivia sculpted by wind erosion.

Wind erosion is a major geomorphological force, especially in arid and semi-arid regions. It is also a major source of land degradation, evaporation, desertification, harmful airborne dust, and crop damage—especially after being increased far above natural rates by human activities such as deforestation, urbanization, and agriculture.

Wind erosion is of two primary varieties: *deflation*, where the wind picks up and carries away loose particles; and *abrasion*, where surfaces are worn down as they are struck by airborne particles carried by wind. Deflation is divided into three categories: (1) *surface creep*, where larger, heavier particles slide or roll along the ground; (2) *saltation*, where particles are lifted a short height into the air, and bounce and saltate across the surface of the soil; and (3) *suspension*, where very small and light particles are lifted into the air by the wind, and are often carried for long distances. Saltation is responsible for the majority (50–70%) of wind erosion, followed by suspension (30–40%), and then surface creep (5–25%). Silty soils tend to be the most affected by wind erosion; silt particles are relatively easily detached and carried away.

Wind erosion is much more severe in arid areas and during times of drought. For example, in the Great Plains, it is estimated that soil loss due to wind erosion can be as much as 6100 times greater in drought years than in wet years.

Mass Movement

Wadi in Makhtesh Ramon, Israel, showing gravity collapse erosion on its banks.

Mass movement is the downward and outward movement of rock and sediments on a sloped surface, mainly due to the force of gravity.

Mass movement is an important part of the erosional process, and is often the first stage in the breakdown and transport of weathered materials in mountainous areas. It moves material from higher elevations to lower elevations where other eroding agents such as streams and glaciers can then pick up the material and move it to even lower elevations. Mass-movement processes are always occurring continuously on all slopes; some mass-movement processes act very slowly; others occur very suddenly, often with disastrous results. Any perceptible down-slope movement of rock or sediment is often referred to in general terms as a landslide. However, landslides can be classified in a much more detailed way that reflects the mechanisms responsible for the movement and the velocity at which the movement occurs. One of the visible topographical manifestations of a very slow form of such activity is a scree slope.

Slumping happens on steep hillsides, occurring along distinct fracture zones, often within materials like clay that, once released, may move quite rapidly downhill. They will often show a spoon-shaped isostatic depression, in which the material has begun to slide downhill. In some cases, the slump is caused by water beneath the slope weakening it. In many cases it is simply the result of poor engineering along highways where it is a regular occurrence.

Surface creep is the slow movement of soil and rock debris by gravity which is usually not perceptible except through extended observation. However, the term can also describe the rolling of dislodged soil particles 0.5 to 1.0 mm (0.02 to 0.04 in) in diameter by wind along the soil surface.

Factors affecting Soil Erosion

Climate

The amount and intensity of precipitation is the main climatic factor governing soil erosion by water. The relationship is particularly strong if heavy rainfall occurs at times when, or in locations where, the soil's surface is not well protected by vegetation. This might be during periods when agricultural activities leave the soil bare, or in semi-arid regions where vegetation is naturally sparse. Wind erosion requires strong winds, particularly during times of drought when vegetation is sparse and soil is dry (and so is more erodible). Other climatic factors such as average temperature and temperature range may also affect erosion, via their effects on vegetation and soil properties. In general, given similar vegetation and ecosystems, areas with more precipitation (especially high-intensity rainfall), more wind, or more storms are expected to have more erosion.

In some areas of the world (e.g. the mid-western USA), rainfall intensity is the primary determinant of erosivity, with higher intensity rainfall generally resulting in more soil erosion by water. The size and velocity of rain drops is also an important factor. Larger and higher-velocity rain drops have greater kinetic energy, and thus their impact will displace soil particles by larger distances than smaller, slower-moving rain drops.

In other regions of the world (e.g. western Europe), runoff and erosion result from relatively low intensities of stratiform rainfall falling onto previously saturated soil. In such situations, rainfall amount rather than intensity is the main factor determining the severity of soil erosion by water.

Soil Structure and Composition

Erosional gully in unconsolidated Dead Sea (Israel) sediments along the southwestern shore. This gully was excavated by floods from the Judean Mountains in less than a year.

The composition, moisture, and compaction of soil are all major factors in determining the erosivity of rainfall. Sediments containing more clay tend to be more resistant to erosion than those with sand or silt, because the clay helps bind soil particles together. Soil containing high levels of organic materials are often more resistant to erosion, because the organic materials coagulate soil colloids and create a stronger, more stable soil structure. The amount of water present in the soil before the precipitation also plays an important role, because it sets limits on the amount of water that can be absorbed by the soil (and hence prevented from flowing on the surface as erosive runoff). Wet, saturated soils will not be able to absorb as much rain water, leading to higher levels of surface runoff and thus higher erosivity for a given volume of rainfall. Soil compaction also affects the permeability of the soil to water, and hence the amount of water that flows away as runoff. More compacted soils will have a larger amount of surface runoff than less compacted soils.

Vegetative Cover

Vegetation acts as an interface between the atmosphere and the soil. It increases the permeability of the soil to rainwater, thus decreasing runoff. It shelters the soil from winds, which results in decreased wind erosion, as well as advantageous changes in microclimate. The roots of the plants bind the soil together, and interweave with other roots, forming a more solid mass that is less susceptible to both water and wind erosion. The removal of vegetation increases the rate of surface erosion.

Topography

The topography of the land determines the velocity at which surface runoff will flow, which in turn determines the erosivity of the runoff. Longer, steeper slopes (especially those without adequate vegetative cover) are more susceptible to very high rates of erosion during heavy rains than shorter, less steep slopes. Steeper terrain is also more prone to mudslides, landslides, and other forms of gravitational erosion processes.

Human Activities that Increase Soil Erosion

Agricultural Practices

Tilled farmland such as this is very susceptible to erosion from rainfall, due to the destruction of vegetative cover and the loosening of the soil during plowing.

Unsustainable agricultural practices are the single greatest contributor to the global increase in erosion rates. The tillage of agricultural lands, which breaks up soil into finer particles, is one of the primary factors. The problem has been exacerbated in modern times, due to mechanized agricultural equipment that allows for deep plowing, which severely increases the amount of soil that is available for transport by water erosion. Others include mono-cropping, farming on steep slopes, pesticide and chemical fertilizer usage (which kill organisms that bind soil together), row-cropping, and the use of surface irrigation. A complex overall situation with respect to defining nutrient losses from soils, could arise as a result of the size selective nature of soil erosion events. Loss of total phosphorus, for instance, in the finer eroded fraction is greater relative to the whole soil. Extrapolating this evidence to predict subsequent behaviour within receiving aquatic systems, the reason is that this more easily transported material may support a lower solution P concentration compared to coarser sized fractions. Tillage also increases wind erosion rates, by dehydrating the soil and breaking it up into smaller particles that can be picked up by the wind. Exacerbating this is the fact that most of the trees are generally removed from agricultural fields, allowing winds to have long, open runs to travel over at higher speeds. Heavy grazing reduces vegetative cover and causes severe soil compaction, both of which increase erosion rates.

Deforestation

In an undisturbed forest, the mineral soil is protected by a layer of *leaf litter* and an *humus* that cover the forest floor. These two layers form a protective mat over the soil that absorbs the impact of rain drops. They are porous and highly permeable to rainfall, and allow rainwater to slow percolate into the soil below, instead of flowing over the surface as runoff. The roots of the trees and plants hold together soil particles, preventing them from being washed away. The vegetative cover acts to reduce the velocity of the raindrops that strike the foliage and stems before hitting the ground, reducing their kinetic energy. However it is the forest floor, more than the canopy, that prevents surface erosion. The terminal velocity of rain drops is reached in about 8 metres (26 feet). Because forest canopies are usually higher than this, rain drops can often regain terminal velocity even after striking the canopy. However, the intact forest floor, with its layers of leaf litter and organic matter, is still able to absorb the impact of the rainfall.

Deforestation causes increased erosion rates due to exposure of mineral soil by removing the humus and litter layers from the soil surface, removing the vegetative cover that binds soil together, and causing heavy soil compaction from logging equipment. Once trees have been removed by fire or logging, infiltration rates become high and erosion low to the degree the forest floor remains intact. Severe fires can lead to significant further erosion if followed by heavy rainfall.

Globally one of the largest contributors to erosive soil loss in the year 2006 is the slash and burn treatment of tropical forests. In a number of regions of the earth, entire sectors of a country have been rendered unproductive. For example, on the Madagascar high central plateau, comprising approximately ten percent of that country's land area, virtually the entire landscape is sterile of vegetation, with gully erosive furrows typically in excess of 50 metres (160 ft) deep and 1 kilometre (0.6 miles) wide. Shifting cultivation is a farming system which sometimes incorporates the slash and burn method in some regions of the world. This degrades the soil and causes the soil to become less and less fertile.

Roads and Urbanization

Urbanization has major effects on erosion processes—first by denuding the land of vegetative cover, altering drainage patterns, and compacting the soil during construction; and next by covering the land in an impermeable layer of asphalt or concrete that increases the amount of surface runoff and increases surface wind speeds. Much of the sediment carried in runoff from urban areas (especially roads) is highly contaminated with fuel, oil, and other chemicals. This increased runoff, in addition to eroding and degrading the land that it flows over, also causes major disruption to surrounding watersheds by altering the volume and rate of water that flows through them, and filling them with chemically polluted sedimentation. The increased flow of water through local waterways also causes a large increase in the rate of bank erosion.

Climate Change

The warmer atmospheric temperatures observed over the past decades are expected to lead to a more vigorous hydrological cycle, including more extreme rainfall events. The rise in sea levels that has occurred as a result of climate change has also greatly increased coastal erosion rates.

Studies on soil erosion suggest that increased rainfall amounts and intensities will lead to greater rates of soil erosion. Thus, if rainfall amounts and intensities increase in many parts of the world as expected, erosion will also increase, unless amelioration measures are taken. Soil erosion rates are expected to change in response to changes in climate for a variety of reasons. The most direct is the change in the erosive power of rainfall. Other reasons include: a) changes in plant canopy caused by shifts in plant biomass production associated with moisture regime; b) changes in litter cover on the ground caused by changes in both plant residue decomposition rates driven by temperature and moisture dependent soil microbial activity as well as plant biomass production rates; c) changes in soil moisture due to shifting precipitation regimes and evapo-transpiration rates, which changes infiltration and runoff ratios; d) soil erodibility changes due to decrease in soil organic matter concentrations in soils that lead to a soil structure that is more susceptible to erosion and increased runoff due to increased soil surface sealing and crusting; e) a shift of winter precipitation from non-erosive snow to erosive rainfall due to increasing winter temperatures; f) melting of permafrost, which induces an erodible soil state from a previously non-erodible one; and g) shifts in land use made necessary to accommodate new climatic regimes.

Studies by Pruski and Nearing indicated that, other factors such as land use unconsidered, it is reasonable to expect approximately a 1.7% change in soil erosion for each 1% change in total precipitation under climate change.

Global Environmental Effects

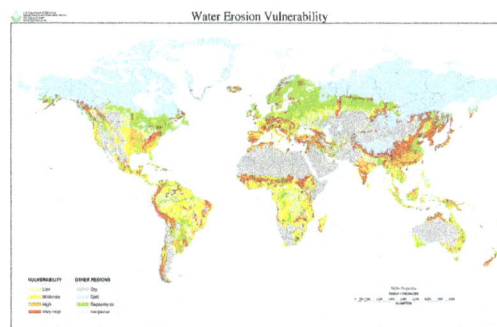

World map indicating areas that are vulnerable to high rates of water erosion.

During the 17th and 18th centuries, Easter Island experienced severe erosion due to deforestation and unsustainable agricultural practices. The resulting loss of topsoil ultimately led to ecological collapse, causing mass starvation and the complete disintegration of the Easter Island civilization.

Due to the severity of its ecological effects, and the scale on which it is occurring, erosion constitutes one of the most significant global environmental problems we face today.

Land Degradation

Water and wind erosion are now the two primary causes of land degradation; combined, they are responsible for 84% of degraded acreage.

Each year, about 75 billion tons of soil is eroded from the land—a rate that is about 13–40 times as fast as the natural rate of erosion. Approximately 40% of the world's agricultural land is seriously degraded. According to the United Nations, an area of fertile soil the size of Ukraine is lost every year because of drought, deforestation and climate change. In Africa, if current trends of soil degradation continue, the continent might be able to feed just 25% of its population by 2025, according to UNU's Ghana-based Institute for Natural Resources in Africa.

The loss of soil fertility due to erosion is further problematic because the response is often to apply chemical fertilizers, which leads to further water and soil pollution, rather than to allow the land to regenerate.

Sedimentation of Aquatic Ecosystems

Soil erosion (especially from agricultural activity) is considered to be the leading global cause of diffuse water pollution, due to the effects of the excess sediments flowing into the world's waterways. The sediments themselves act as pollutants, as well as being carriers for other pollutants, such as attached pesticide molecules or heavy metals.

The effect of increased sediments loads on aquatic ecosystems can be catastrophic. Silt can smother the spawning beds of fish, by filling in the space between gravel on the stream bed. It also reduces their food supply, and causes major respiratory issues for them as sediment enters their gills. The biodiversity of aquatic plant and algal life is reduced, and invertebrates are also unable to survive and reproduce. While the sedimentation event itself might be relatively short-lived, the ecological disruption caused by the mass die off often persists long into the future.

One of the most serious and long-running water erosion problems worldwide is in the People's Republic of China, on the middle reaches of the Yellow River and the upper reaches of the Yangtze River. From the Yellow River, over 1.6 billion tons of sediment flows into the ocean each year. The sediment originates primarily from water erosion in the Loess Plateau region of the northwest.

Airborne Dust Pollution

Soil particles picked up during wind erosion of soil are a major source of air pollution, in the form of airborne particulates—"dust". These airborne soil particles are often contaminated with toxic chemicals such as pesticides or petroleum fuels, posing ecological and public health hazards when they later land, or are inhaled/ingested.

Dust from erosion acts to suppress rainfall and changes the sky color from blue to white, which leads to an increase in red sunsets. Dust events have been linked to a decline in the health of coral reefs across the Caribbean and Florida, primarily since the 1970s. Similar dust plumes originate in the Gobi desert, which combined with pollutants, spread large distances downwind, or eastward, into North America.

Monitoring, Measuring and Modeling Soil Erosion

Terracing is an ancient technique that can significantly slow the rate of water erosion on cultivated slopes.

Monitoring and modeling of erosion processes can help people better understand the causes of soil erosion, make predictions of erosion under a range of possible conditions, and plan the implementation of preventative and restorative strategies for erosion. However, the complexity of erosion processes and the number of scientific disciplines that must be considered to understand and model them (e.g. climatology, hydrology, geology, soil science, agriculture, chemistry, physics, etc.) makes accurate modelling challenging. Erosion models are also non-linear, which makes them difficult to work with numerically, and makes it difficult or impossible to scale up to making predictions about large areas from data collected by sampling smaller plots.

The most commonly used model for predicting soil loss from water erosion is the *Universal Soil Loss Equation (USLE)*. This was developed in the 1960s and 1970s. It estimates the average annual soil loss A on a plot-sized area as:

$$A = RKLSCP$$

where R is the rainfall erosivity factor, K is the soil erodibility factor, L and S are topographic factors representing length and slope, C is the cover and management factor and P is the support practices factor.

Despite the USLE's plot-scale spatial focus, the model has often been used to estimate soil erosion on much larger areas, such as watersheds or even whole continents. For example, RUSLE has recently been used to quantify soil erosion across the whole of Europe . This is scientifically controversial, for several reasons. One major problem is that the USLE cannot simulate gully erosion, and so erosion from gullies is ignored in any USLE-based assessment of erosion. Yet erosion from gullies can be a substantial proportion (10–80%) of total erosion on cultivated and grazed land.

During the 50 years since the introduction of the USLE, many other soil erosion models have been developed. But because of the complexity of soil erosion and its constituent processes, all erosion models can give unsatisfactory results when validated i.e. when model predictions are compared with real-world measurements of erosion. Thus new soil erosion models continue to be developed.

Some of these remain USLE-based, e.g. the G2 model . Other soil erosion models have largely (e.g. the Water Erosion Prediction Project model) or wholly (e.g. the Rangeland Hydrology and Erosion Model) abandoned usage of USLE elements.

Prevention and Remediation

A windbreak (the row of trees) planted next to an agricultural field, acting as a shield against strong winds. This reduces the effects of wind erosion, and provides many other benefits.

The most effective known method for erosion prevention is to increase vegetative cover on the land, which helps prevent both wind and water erosion. Terracing is an extremely effective means of erosion control, which has been practiced for thousands of years by people all over the world. Windbreaks (also called shelterbelts) are rows of trees and shrubs that are planted along the edges of agricultural fields, to shield the fields against winds. In addition to significantly reducing wind erosion, windbreaks provide many other benefits such as improved microclimates for crops (which are sheltered from the dehydrating and otherwise damaging effects of wind), habitat for beneficial bird species, carbon sequestration, and aesthetic improvements to the agricultural landscape. Traditional planting methods, such as mixed-cropping (instead of monocropping) and crop rotation have also been shown to significantly reduce erosion rates. Crop residues play a role in the mitigation of erosion, because they reduce the impact of raindrops breaking up the soil particles. There is a higher potential for erosion when producing potatoes than when growing cereals, or oilseed crops. Forages have a fibrous root system, which helps combat erosion by anchoring the plants to the top layer of the soil, and covering the entirety of the field, as it is a non-row crop. In tropical coastal systems, properties of mangroves have been examined as a potential means to reduce soil erosion. Their complex root structures are known to help reduce wave damage from storms and flood impacts while binding and building soils. These roots can slow down water flow, leading to the deposition of sediments and reduced erosion rates. However, in order to maintain sediment balance, adequate mangrove forest width needs to be present.

Drought

A drought is a period of below-average precipitation in a given region, resulting in prolonged shortages in its water supply, whether atmospheric, surface water or ground water. A drought can last for months or years, or may be declared after as few as 15 days. It can have a substantial impact on the ecosystem and agriculture of the affected region and harm to the local economy. Annual dry seasons in the tropics significantly increase the chances of a drought developing and subsequent bush fires. Periods of heat can significantly worsen drought conditions by hastening evaporation of water vapour.

Many plant species, such as those in the family Cactaceae (or cacti), have drought tolerance adaptations like reduced leaf area and waxy cuticles to enhance their ability to tolerate drought. Some others survive dry periods as buried seeds. Semi-permanent drought produces arid biomes such as deserts and grasslands. Prolonged droughts have caused mass migrations and humanitarian crises. Most arid ecosystems have inherently low productivity. The most prolonged drought ever in the world in recorded history occurred in the Atacama Desert in Chile (400 Years).

Contraction/Desiccation cracks in dry earth (Sonoran desert, Mexico).

Causes of Drought

Precipitation Deficiency

Mechanisms of producing precipitation include convective, stratiform, and orographic rainfall. Convective processes involve strong vertical motions that can cause the overturning of the atmosphere in that location within an hour and cause heavy precipitation, while stratiform processes involve weaker upward motions and less intense precipitation over a longer duration. Precipitation can be divided into three categories, based on whether it falls as liquid water, liquid water that freezes on contact with the surface, or ice. Droughts occur mainly in areas where normal levels of rainfall are, in themselves, low. If these factors do not support precipitation volumes sufficient to reach the surface over a sufficient time, the result is a drought. Drought can be triggered by a high level of reflected sunlight and above average prevalence of high pressure systems, winds carrying continental, rather than oceanic air masses, and ridges of high pressure areas aloft can prevent or restrict the developing of thunderstorm activity or rainfall over one certain region. Once a region is within drought, feedback mechanisms such as local arid air, hot conditions which can promote warm core ridging, and minimal evapotranspiration can worsen drought conditions.

Dry Season

Within the tropics, distinct, wet and dry seasons emerge due to the movement of the Intertropical Convergence Zone or Monsoon trough. The dry season greatly increases drought occurrence, and is characterized by its low humidity, with watering holes and rivers drying up. Because of the lack of these watering holes, many grazing animals are forced to migrate due to the lack of water and feed to more fertile spots. Examples of such animals are zebras, elephants, and wildebeest. Because of the lack of water in the plants, bushfires are common. Since water vapor becomes more energetic with increasing temperature, more water vapor is required to increase relative humidity values to 100% at higher temperatures (or to get the temperature to fall to the dew point). Periods of warmth quicken the pace of fruit and vegetable production, increase evaporation and transpiration from plants, and worsen drought conditions.

Sheep on a drought affected paddock near Uranquinty, New South Wales.

El Niño

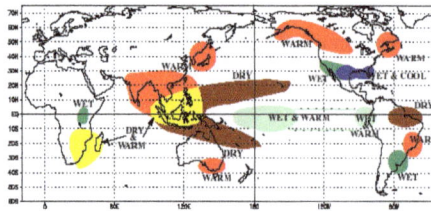

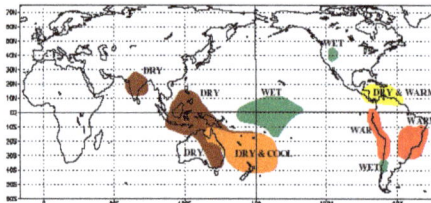

Regional impacts of warm ENSO episodes (El Niño)

Drier and hotter weather occurs in parts of the Amazon River Basin, Colombia, and Central America during El Niño events. Winters during the El Niño are warmer and drier than average conditions in the Northwest, northern Midwest, and northern Mideast United States, so those regions experience reduced snowfalls. Conditions are also drier than normal from December to February in south-central Africa, mainly in Zambia, Zimbabwe, Mozambique, and Botswana. Direct effects of El Niño resulting in drier conditions occur in parts of Southeast Asia and Northern Australia, increasing bush fires, worsening haze, and decreasing air quality dramatically. Drier-than-normal conditions are also in general observed in Queensland, inland Victoria, inland New South Wales, and eastern Tasmania from June to August. As warm water spreads from the west Pacific and the Indian Ocean to the east Pacific, it causes extensive drought in the western Pacific. Singapore experienced the driest February in 2014 since records began in 1869, with only 6.3 mm of rain falling in the month and temperatures hitting as high as 35 °C on 26 February. The years 1968 and 2005 had the next driest Februaries, when 8.4 mm of rain fell.

Erosion and Human Activities

Fires on Borneo and Sumatra, 2006. People use slash-and-burn deforestation to clear land for agriculture.

Human activity can directly trigger exacerbating factors such as over farming, excessive irrigation, deforestation, and erosion adversely impact the ability of the land to capture and hold water. In arid climates, the main source of erosion is wind. Erosion can be the result of material movement by the wind. The wind can cause small particles to be lifted and therefore moved to another region (deflation). Suspended particles within the wind may impact on solid objects causing erosion by abrasion (ecological succession). Wind erosion generally occurs in areas with little or no vegetation, often in areas where there is insufficient rainfall to support vegetation.

Fields outside Benambra, Victoria, Australia suffering from drought conditions.

Loess is a homogeneous, typically nonstratified, porous, friable, slightly coherent, often calcareous, fine-grained, silty, pale yellow or buff, windblown (Aeolian) sediment. It generally occurs as a widespread blanket deposit that covers areas of hundreds of square kilometers and tens of meters thick. Loess often stands in either steep or vertical faces. Loess tends to develop into highly rich soils. Under appropriate climatic conditions, areas with loess are among the most agriculturally productive in the world. Loess deposits are geologically unstable by nature, and will erode very readily. Therefore, windbreaks (such as big trees and bushes) are often planted by farmers to reduce the wind erosion of loess. Wind erosion is much more severe in arid areas and during times of drought. For example, in the Great Plains, it is estimated that soil loss due to wind erosion can be as much as 6100 times greater in drought years than in wet years.

Climate Change

Activities resulting in global climate change are expected to trigger droughts with a substantial impact on agriculture throughout the world, and especially in developing nations. Overall, global warming will result in increased world rainfall. Along with drought in some areas, flooding and erosion will increase in others. Paradoxically, some proposed solutions to global warming that focus on more active techniques, solar radiation management through the use of a space sunshade for one, may also carry with them increased chances of drought.

Types

As a drought persists, the conditions surrounding it gradually worsen and its impact on the local population gradually increases. People tend to define droughts in three main ways:

1. Meteorological drought is brought about when there is a prolonged time with less than average precipitation. Meteorological drought usually precedes the other kinds of drought.

2. Agricultural droughts are droughts that affect crop production or the ecology of the range. This condition can also arise independently from any change in precipitation levels when soil conditions and erosion triggered by poorly planned agricultural endeavors cause a shortfall in water available to the crops. However, in a traditional drought, it is caused by an extended period of below average precipitation.

3. Hydrological drought is brought about when the water reserves available in sources such as aquifers, lakes and reservoirs fall below the statistical average. Hydrological drought tends to show up more slowly because it involves stored water that is used but not replenished. Like an agricultural drought, this can be triggered by more than just a loss of rainfall. For instance, Kazakhstan was recently awarded a large amount of money by the World Bank to restore water that had been diverted to other nations from the Aral Sea under Soviet rule. Similar circumstances also place their largest lake, Balkhash, at risk of completely drying out.

Consequences of Drought

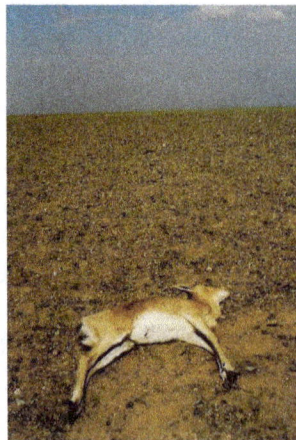

A Mongolian gazelle dead due to drought.

The effects of droughts and water shortages can be divided into three groups: environmental, economic and social consequences. In the case of environmental effects: lower surface and subterranean water levels, lower flow levels (with a decrease below the minimum leading to direct danger for amphibian life), increased pollution of surface water, the drying out of wetlands, more and larger fires, higher deflation intensity, losing biodiversity, worse health of trees and the appearance of pests and den droid diseases. Economic losses include lower agricultural, forests, game and fishing output, higher food production costs, lower energy production levels in hydro plants, losses caused by depleted water tourism and transport revenue, problems with water supply for the energy sector and technological processes in metallurgy, mining, the chemical, paper, wood, foodstuff industries etc., disruption of water supplies for municipal economies. Meanwhile, social costs include the negative effect on the health of people directly exposed to this phenomenon (excessive heat waves), possible limitation of water supplies and its increased pollution levels, high food costs, stress caused by failed harvests, etc. This is why droughts and fresh water shortages may be considered as a factor which increases the gap between developed and developing countries.

The effect varies according to vulnerability. For example, subsistence farmers are more likely to migrate during drought because they do not have alternative food sources. Areas with populations that depend on water sources as a major food source are more vulnerable to famine.

Drought can also reduce water quality, because lower water flows reduce dilution of pollutants and increase contamination of remaining water sources. Common consequences of drought include:

- Diminished crop growth or yield productions and carrying capacity for livestock
- Dust bowls, themselves a sign of erosion, which further erode the landscape
- Dust storms, when drought hits an area suffering from desertification and erosion
- Famine due to lack of water for irrigation
- Habitat damage, affecting both terrestrial and aquatic wildlife
- Hunger, drought provides too little water to support food crops.
- Malnutrition, dehydration and related diseases
- Mass migration, resulting in internal displacement and international refugees
- Reduced electricity production due to reduced water flow through hydroelectric dams
- Shortages of water for industrial users
- Snake migration, which results in snakebites
- Social unrest
- War over natural resources, including water and food
- Wildfires, such as Australian bushfires, are more common during times of drought and even death of people.
- Exposure and oxidation of acid sulfate soils due to falling surface and groundwater levels.
- Cyanotoxin accumulation within food chains and water supply, some of which are among the most potent toxins known to science, can cause cancer with low exposure over long term. High levels of microcystin has been found in San Francisco Bay Area salt water shellfish and fresh water supplies throughout the state of California in 2016.

Globally

Drought is a normal, recurring feature of the climate in most parts of the world. It is among the earliest documented climatic events, present in the Epic of Gilgamesh and tied to the biblical story of Joseph's arrival in and the later Exodus from Ancient Egypt. Hunter-gatherer migrations in 9,500 BC Chile have been linked to the phenomenon, as has the exodus of early humans out of Africa and into the rest of the world around 135,000 years ago.

A South Dakota farm during the Dust Bowl, 1936

Examples

Well-known historical droughts include:

- 1900 India killing between 250,000 and 3.25 million.

- 1921–22 Soviet Union in which over 5 million perished from starvation due to drought

- 1928–30 Northwest China resulting in over 3 million deaths by famine.

- 1936 and 1941 Sichuan Province China resulting in 5 million and 2.5 million deaths respectively.

- The 1997–2009 Millennium Drought in Australia led to a water supply crisis across much of the country. As a result, many desalination plants were built for the first time.

- In 2006, Sichuan Province China experienced its worst drought in modern times with nearly 8 million people and over 7 million cattle facing water shortages.

- 12-year drought that was devastating southwest Western Australia, southeast South Australia, Victoria and northern Tasmania was "very severe and without historical precedent".

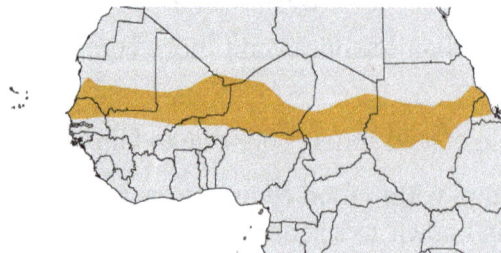

Affected areas in the western Sahel belt during the 2012 drought.

The Darfur conflict in Sudan, also affecting Chad, was fueled by decades of drought; combination of drought, desertification and overpopulation are among the causes of the Darfur conflict, because the Arab Baggara nomads searching for water have to take their livestock further south, to land mainly occupied by non-Arab farming people.

Approximately 2.4 billion people live in the drainage basin of the Himalayan rivers. India, China, Pakistan, Bangladesh, Nepal and Myanmar could experience floods followed by droughts in coming decades. Drought in India affecting the Ganges is of particular concern, as it provides drinking water and agricultural irrigation for more than 500 million people. The west coast of North America, which gets much of its water from glaciers in mountain ranges such as the Rocky Mountains and Sierra Nevada, also would be affected.

Drought affected area in Karnataka, India in 2012.

In 2005, parts of the Amazon basin experienced the worst drought in 100 years. A 23 July 2006 article reported Woods Hole Research Center results showing that the forest in its present form could survive only three years of drought. Scientists at the Brazilian National Institute of Amazonian Research argue in the article that this drought response, coupled with the effects of deforestation on regional climate, are pushing the rainforest towards a "tipping point" where it would irreversibly start to die. It concludes that the rainforest is on the brink of being turned into savanna or desert, with catastrophic consequences for the world's climate. According to the WWF, the combination of climate change and deforestation increases the drying effect of dead trees that fuels forest fires.

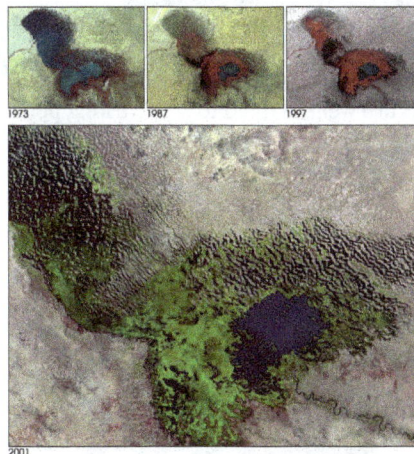

Lake Chad in a 2001 satellite image. The lake has shrunk by 95% since the 1960s.

By far the largest part of Australia is desert or semi-arid lands commonly known as the outback. A 2005 study by Australian and American researchers investigated the desertification of the interior, and suggested that one explanation was related to human settlers who arrived about 50,000 years ago. Regular burning by these settlers could have prevented monsoons from reaching interior Australia. In June 2008 it became known that an expert panel had warned of long term, maybe irreversible, severe ecological damage for the whole Murray-Darling basin if it did not receive sufficient water by October 2008. Australia could experience more severe droughts and they could become more frequent in the future, a government-commissioned report said on July 6, 2008. Australian environmentalist Tim Flannery, predicted that unless it made drastic changes, Perth in Western Australia could become the world's first ghost metropolis, an abandoned city with no more water to sustain its population. The long Australian Millennial drought broke in 2010.

Recurring droughts leading to desertification in East Africa have created grave ecological catastrophes, prompting food shortages in 1984–85, 2006 and 2011. During the 2011 drought, an estimated 50,000 to 150,000 people were reported to have died, though these figures and the extent of the crisis are disputed. In February 2012, the UN announced that the crisis was over due to a scaling up of relief efforts and a bumper harvest. Aid agencies subsequently shifted their emphasis to recovery efforts, including digging irrigation canals and distributing plant seeds.

In 2012, a severe drought struck the western Sahel. The Methodist Relief & Development Fund (MRDF) reported that more than 10 million people in the region were at risk of famine due to a month-long heat wave that was hovering over Niger, Mali, Mauritania and Burkina Faso. A fund of about £20,000 was distributed to the drought-hit countries.

Protection, Mitigation and Relief

Succulent plants are well-adapted to survive long periods of drought.

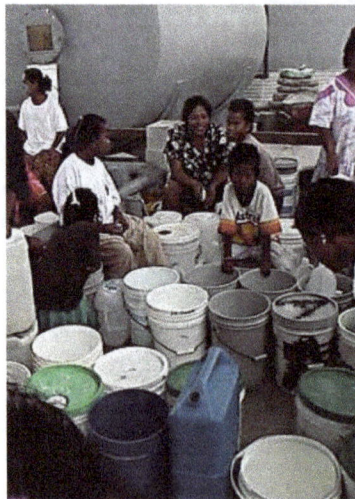

Water distribution on Marshall Islands during El Niño.

Agriculturally, people can effectively mitigate much of the impact of drought through irrigation and crop rotation. Failure to develop adequate drought mitigation strategies carries a grave human cost in the modern era, exacerbated by ever-increasing population densities. President Roosevelt on April 27, 1935, signed documents creating the Soil Conservation Service (SCS)—now the Natural Resources Conservation Service (NRCS). Models of the law were sent to each state where they were enacted. These were the first enduring practical programs to curtail future susceptibility to drought, creating agencies that first began to stress soil conservation measures to protect farm lands today. It was not until the 1950s that there was an importance placed on water conservation was put into the existing laws (NRCS 2014).

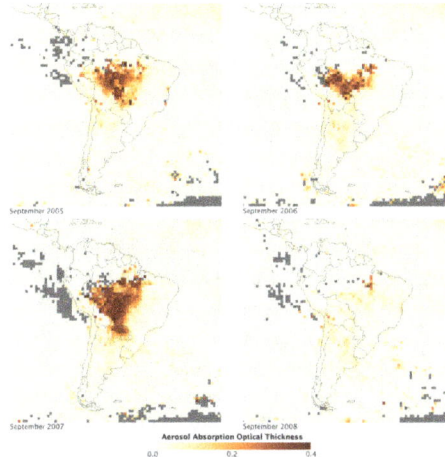

Aerosols over the Amazon each September for four burning seasons (2005 through 2008) during the Amazon basin drought. The aerosol scale (yellow to dark reddish-brown) indicates the relative amount of particles that absorb sunlight.

Strategies for drought protection, mitigation or relief include:

- Dams - many dams and their associated reservoirs supply additional water in times of drought.

- Cloud seeding - a form of intentional weather modification to induce rainfall. This remains a hotly debated topic, as the United States National Research Council released a report in 2004 stating that to date, there is still no convincing scientific proof of the efficacy of intentional weather modification.

- Desalination - of sea water for irrigation or consumption.

- Drought monitoring - Continuous observation of rainfall levels and comparisons with current usage levels can help prevent man-made drought. For instance, analysis of water usage in Yemen has revealed that their water table (underground water level) is put at grave risk by over-use to fertilize their Khat crop. Careful monitoring of moisture levels can also help predict increased risk for wildfires, using such metrics as the Keetch-Byram Drought Index or Palmer Drought Index.

- Land use - Carefully planned crop rotation can help to minimize erosion and allow farmers to plant less water-dependent crops in drier years.

- Outdoor water-use restriction - Regulating the use of sprinklers, hoses or buckets on outdoor plants, filling pools, and other water-intensive home maintenance tasks. Xeriscaping yards can significantly reduce unnecessary water use by residents of towns and cities.

- Rainwater harvesting - Collection and storage of rainwater from roofs or other suitable catchments.

- Recycled water - Former wastewater (sewage) that has been treated and purified for reuse.

- Transvasement - Building canals or redirecting rivers as massive attempts at irrigation in drought-prone areas.

Flash Flood

An urban underpass during normal conditions (upper) and after fifteen minutes of heavy rain (lower)

Driving through a flash-flooded road

A flash flood after a thunderstorm in the Gobi, Mongolia

A flash flood is a rapid flooding of geomorphic low-lying areas: washes, rivers, dry lakes and basins. It may be caused by heavy rain associated with a severe thunderstorm, hurricane, tropical storm, or meltwater from ice or snow flowing over ice sheets or snowfields. Flash floods may occur after the collapse of a natural ice or debris dam, or a human structure such as a man-made dam, as occurred before the Johnstown Flood of 1889. Flash floods are distinguished from regular floods by a timescale of less than six hours. The water that is temporarily available is often used by foliage with rapid germination and short growth cycles, and by specially adapted animal life.

Causes

Flash floods can occur under several types of conditions. Flash flooding occurs when it rains rapidly on saturated soil or dry soil that has poor absorption ability. The runoff collects in gullies and streams and, as they join to form larger volumes, often forms a fast flowing front of water and debris.

Flash floods most often occur in normally dry areas that have recently received precipitation, but they may be seen anywhere downstream from the source of the precipitation, even many miles from the source. In areas on or near volcanoes, flash floods have also occurred after eruptions, when glaciers have been melted by the intense heat. Flash floods are known to occur in the highest mountain ranges of the United States and are also common in the arid plains of the Southwestern United States. Flash flooding can also be caused by extensive rainfall released by hurricanes and other tropical storms, as well as the sudden thawing effect of ice dams. Human activities can also cause flash floods to occur. When dams fail, a large quantity of water can be released and destroy everything in its path.

Hazards

The United States National Weather Service gives the advice "Turn Around, Don't Drown" for flash floods; that is, it recommends that people get out of the area of a flash flood, rather than trying to cross it. Many people tend to underestimate the dangers of flash floods. What makes flash floods most dangerous is their sudden nature and fast-moving water. A vehicle provides little to no protection against being swept away; it may make people overconfident and less likely to avoid the flash flood. More than half of the fatalities attributed to flash floods are people swept away in vehicles when trying to cross flooded intersections. As little as 2 feet (0.61 m) of water is enough to carry away most SUV-sized vehicles. The U.S. National Weather Service reported in 2005 that, using a national 30-year average, more people die yearly in floods, 127 on average, than by lightning (73), tornadoes (65), or hurricanes (16).

In deserts, flash floods can be particularly deadly for several reasons. First, storms in arid regions are infrequent, but they can deliver an enormous amount of water in a very short time. Second, these rains often fall on poorly absorbent and often clay-like soil, which greatly increase the amount of runoff that rivers and other water channels have to handle. These regions tend not to have the infrastructure that wetter regions have to divert water from structures and roads, such as storm drains, culverts, and retention basins, either because of sparse population, poverty, or because residents believe the risk of flash floods is not high enough to justify the expense. In fact, in some areas, desert roads frequently cross dry river and creek beds without bridges. From the driver's perspective, there may be clear weather, when a river unexpectedly forms ahead of or around the vehicle in a matter of seconds. Finally, the lack of regular rain to clear water channels may cause flash floods in deserts to be headed by large amounts of debris, such as rocks, branches, and logs.

Deep slot canyons can be especially dangerous to hikers as they may be flooded by a storm that occurs on a mesa miles away, sweeps through the canyon, and makes it difficult to climb up and out of the way to avoid the flood.

Desertification

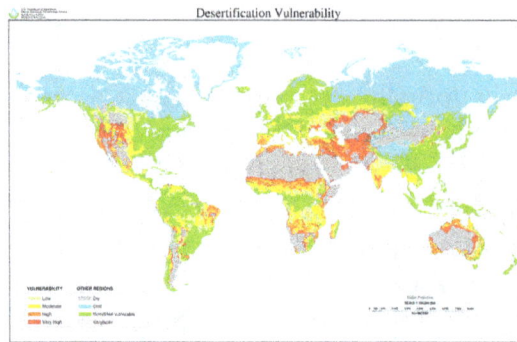

Global desertification vulnerability map

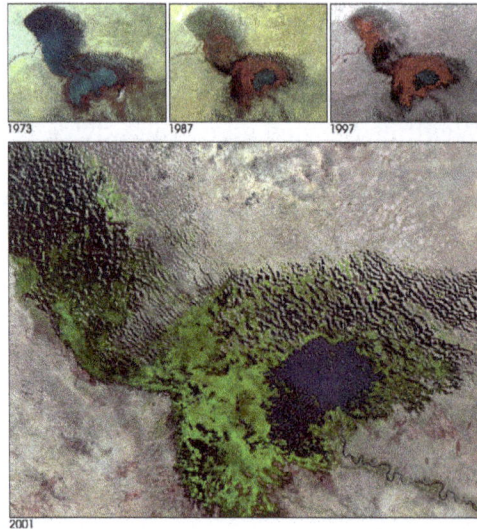

Lake Chad in a 2001 satellite image, with the actual lake in blue. The lake has shrunk by 94% since the 1960s.

Desertification is a type of land degradation in which relatively dry area of land becomes increasingly arid, typically losing its bodies of water as well as vegetation and wildlife. It is caused by a variety of factors, such as through climate change and through the overexploitation of soil through human activity. When deserts appear automatically over the natural course of a planet's life cycle, then it can be called a natural phenomenon; however, when deserts emerge due to the rampant and unchecked depletion of nutrients in soil that are essential for it to remain arable, then a virtual "soil death" can be spoken of, which traces its cause back to human overexploitation. Desertification is a significant global ecological and environmental problem.

Definitions

Considerable controversy exists over the proper definition of the term "desertification" for which Helmut Geist (2005) has identified more than 100 formal definitions. The most widely accepted of these is that of the Princeton University Dictionary which defines it as "the process of fertile land transforming into desert typically as a result of deforestation, drought or improper/inappropriate agriculture". Desertification has been neatly defined in the text of the United Nations Convention to Combat Desertification (UNCCD) as "land degradation in arid, semi-arid and dry sub-humid regions resulting from various factors, including climatic variations and human activities."

Another major contribution to the controversy comes from the sub-grouping of types of desertification. Spanning from the very vague yet shortsighted view as the "man-made-desert" to the more broad yet less focused type as the "Non-pattern-Desert"

The earliest known discussion of the topic arose soon after the French colonization of West Africa, when the Comité d'Etudes commissioned a study on *desséchement progressif* to explore the prehistoric expansion of the Sahara Desert.

History

The world's most noted deserts have been formed by natural processes interacting over long intervals of time. During most of these times, deserts have grown and shrunk independent of human activities. Paleodeserts are large sand seas now inactive because they are stabilized by vegetation, some extending beyond the present margins of core deserts, such as the Sahara, the largest hot desert.

Desertification has played a significant role in human history, contributing to the collapse of several large empires, such as Carthage, Greece, and the Roman Empire, as well as causing displacement of local populations. Historical evidence shows that the serious and extensive land deterioration occurring several centuries ago in arid regions had three epicenters: the Mediterranean, the Mesopotamian Valley, and the Loess Plateau of China, where population was dense.

Areas Affected

Sun, Moon, and large telescopes above the Desert

Drylands occupy approximately 40–41% of Earth's land area and are home to more than 2 billion people. It has been estimated that some 10–20% of drylands are already degraded, the total area affected by desertification being between 6 and 12 million square kilometres, that about 1–6% of the inhabitants of drylands live in desertified areas, and that a billion people are under threat from further desertification.

As of 1998, the then-current degree of southward expansion of the Sahara was not well known, due to a lack of recent, measurable expansion of the desert into the Sahel at the time.

Causes of Desertification in Sahel:

The impact of global warming and human activities are presented in the Sahel. In this area, the level of desertification is very high compared to other areas in the world.

All areas situated in the eastern part of Africa (i.e. in the Sahel region) are characterized by a dry climate, hot temperatures, and low rainfall (300–750 mm rainfall per year). So, droughts are the rule in the Sahel region.

Development of the desertification process in Sahel:

Some studies have shown that Africa has lost approximately 650 000 km² of its productive agricultural land over the past 50 years. The propagation of desertification in this area is considerable.

Some statistics have shown that since 1900, the Sahara has expanded by 250 km, covering an additional area of 6000 square kilometers.

Impacts of Desertification in Sahel:

The survey, done by the research institute for development, had demonstrated that this means dryness is spreading fast in the Sahelian countries. Desertification in the Sahel can affect more than one billion of its inhabitants. 70% of the arid area has deteriorated and water resources have disappeared, leading to soil degradation. The loss of topsoil means that plants cannot take root firmly and can be uprooted by torrential water or strong winds.

The United Nations Convention (UNC) says that about six million Sahelian citizens would have to give up the desertified zones of sub-Saharan Africa for North Africa and Europe between 1997 and 2020.

China and the Gobi

Another major area that is being impacted by desertification is the Gobi Desert. Currently, the Gobi desert is the fastest moving desert on Earth; according to some researchers, the Gobi Desert swallows up over 1,300 miles of land annually. This has destroyed many villages in its path. Currently, photos show that the Gobi Desert has expanded to the point the entire nation of Croatia could fit inside its area. This is causing a major problem for the people of China. They will soon have to deal with the desert as it creeps closer. Although the Gobi Desert itself is still a distance away from Beijing, reports from field studies state there are large sand dunes forming only 70km(43.5m) outside of the city.

Vegetation Patterning

As the desertification takes place, the landscape may progress through different stages and continuously transform in appearance. On gradually sloped terrain, desertification can create increasingly larger empty spaces over a large strip of land, a phenomenon known as "Brousse tigrée". A mathematical model of this phenomenon proposed by C. Klausmeier attributes this patterning to dynamics in plant-water interaction. One outcome of this observation suggests an optimal planting strategy for agriculture in arid environments.

Causes

Preventing Man-made Overgrazing

The immediate cause is the loss of most vegetation. This is driven by a number of factors, alone or in combination, such as drought, climatic shifts, tillage for agriculture, overgrazing and deforestation for fuel or construction materials. Vegetation plays a major role in determining the biological composition of the soil. Studies have shown that, in many environments, the rate of erosion and runoff decreases exponentially with increased vegetation cover. Unprotected, dry soil surfaces blow away with the wind or are washed away by flash floods, leaving infertile lower soil layers that bake in the sun and become an unproductive hardpan. Controversially, Allan Savory has claimed that the controlled movement of herds of livestock, mimicking herds of grazing wildlife, can reverse desertification.

Goats inside of a pen in Norte Chico, Chile. Overgrazing of drylands by poorly managed traditional herding is one of the primary causes of desertification.

Wildebeest in Masai Mara during the Great Migration. Overgrazing is not caused by nomadic grazers in huge populations of travel herds, nor by holistic planned grazing.

A shepherd guiding his sheep through the high desert outside of Marrakech, Morocco

Poverty

At least 90% of the inhabitants of drylands live in developing nations, where they also suffer from poor economic and social conditions. This situation is exacerbated by land degradation because of the reduction in productivity, the precariousness of living conditions and the difficulty of access to resources and opportunities.

A downward spiral is created in many underdeveloped countries by overgrazing, land exhaustion and overdrafting of groundwater in many of the marginally productive world regions due to over-population pressures to exploit marginal drylands for farming. Decision-makers are understand-ably averse to invest in arid zones with low potential. This absence of investment contributes to the marginalisation of these zones. When unfavourable agro-climatic conditions are combined with an absence of infrastructure and access to markets, as well as poorly adapted production techniques and an underfed and undereducated population, most such zones are excluded from development.

Desertification often causes rural lands to become unable to support the same sized populations that previously lived there. This results in mass migrations out of rural areas and into urban areas, particularly in Africa. These migrations into the cities often cause large numbers of unemployed people, who end up living in slums.

Countermeasures and Prevention

Anti-sand shields in north Sahara, Tunisia

Jojoba plantations, such as those shown, have played a role in combating edge effects of desertification in the Thar Desert, India.

Techniques and countermeasures exist for mitigating or reversing the effects of desertification, and some possess varying levels of difficulty. For some, there are numerous barriers to their imple-mentation. Yet for others, the solution simply requires the exercise of human reason.

One less difficult solution that has been proposed, however controversial it may be, is to bring about a cap on the population growth, and in fact to turn this into a population decay, so that each year there will gradually exist fewer and fewer humans who require the land to be depleted even further in order to grow their food.

One proposed barrier is that the costs of adopting sustainable agricultural practices sometimes exceed the benefits for individual farmers, even while they are socially and environmentally bene-ficial. Another issue is a lack of political will, and lack of funding to support land reclamation and anti-desertification programs.

Desertification is recognized as a major threat to biodiversity. Some countries have developed Bio-diversity Action Plans to counter its effects, particularly in relation to the protection of endangered flora and fauna.

Reforestation gets at one of the root causes of desertification and is not just a treatment of the symptoms. Environmental organizations work in places where deforestation and desertification are contributing to extreme poverty. There they focus primarily on educating the local population about the dangers of deforestation and sometimes employ them to grow seedlings, which they transfer to severely deforested areas during the rainy season. The Food and Agriculture Organization of the United Nations launched the FAO Drylands Restoration Initiative in 2012 to draw together knowledge and experience on dryland restoration. In 2015, FAO published global guidelines for the restoration of degraded forests and landscapes in drylands, in collaboration with the Turkish Ministry of Forestry and Water Affairs and the Turkish Cooperation and Coordination Agency.

Currently, one of the major methods that has been finding success in this battle with desertification. This is known as China's "Great Green Wall." This wall is a much larger scaled version of what American farmers did in the 1930's to stop the great Midwest dust bowl. This plan was proposed in the late 70's, and has become a major ecological engineering project that is not predicted to end until the year 2055. According to Chinese reports, there have been nearly 66,000,000,000 tress planted in China's great green wall. Due to the success that China has been finding in stopping the spread of desertification. Through their success with their wall, plans are currently be made in Africa to start a "wall" along the borders of the Sahara desert as well.

Techniques focus on two aspects: provisioning of water, and fixation and hyper-fertilizing soil.

Fixating the soil is often done through the use of shelter belts, woodlots and windbreaks. Windbreaks are made from trees and bushes and are used to reduce soil erosion and evapotranspiration. They were widely encouraged by development agencies from the middle of the 1980s in the Sahel area of Africa.

Some soils (for example, clay), due to lack of water can become consolidated rather than porous (as in the case of sandy soils). Some techniques as zaï or tillage are then used to still allow the planting of crops.

Another technique that is useful is contour trenching. This involves the digging of 150m long, 1m deep trenches in the soil. The trenches are made parallel to the height lines of the landscape, preventing the water from flowing within the trenches and causing erosion. Stone walls are placed around the trenches to prevent the trenches from closing up again. The method was invented by Peter Westerveld.

Enriching of the soil and restoration of its fertility is often done by plants. Of these, leguminous plants which extract nitrogen from the air and fix it in the soil, and food crops/trees as grains, barley, beans and dates are the most important. Sand fences can also be used to control drifting of soil and sand erosion.

Some research centra (such as Bel-Air Research Center IRD/ISRA/UCAD) are also experimenting with the inoculation of tree species with mycorrhiza in arid zones. The mycorrhiza are basically fungi attaching themselves to the roots of the plants. They hereby create a symbiotic relation with the trees, increasing the surface area of the tree's roots greatly (allowing the tree to gather much more nutrients from the soil).

As there are many different types of deserts, there are also different types of desert reclamation methodologies. An example for this is the salt-flats in the Rub' al Khali desert in Saudi-Arabia. These salt-flats are one of the most promising desert areas for seawater agriculture and could be revitalized without the use of freshwater or much energy.

Farmer-managed natural regeneration (FMNR) is another technique that has produced successful results for desert reclamation. Since 1980, this method to reforest degraded landscape has been applied with some success in Niger. This simple and low-cost method has enabled farmers to regenerate some 30,000 square kilometers in Niger. The process involves enabling native sprouting tree growth through selective pruning of shrub shoots. The residue from pruned trees can be used to provide mulching for fields thus increasing soil water retention and reducing evaporation. Additionally, properly spaced and pruned trees can increase crop yields. The Humbo Assisted Regeneration Project which uses FMNR techniques in Ethiopia has received money from The World Bank's BioCarbon Fund, which supports projects that sequester or conserve carbon in forests or agricultural ecosystems.

It is argued that managed grazing methods are able to restore grasslands.

Managed Grazing

Restoring grasslands store CO_2 from the air into plant material. Grazing livestock, usually not left to wander, would eat the grass and would minimize any grass growth while grass left alone would eventually grow to cover its own growing buds, preventing them from photosynthesizing and killing the plant. A method proposed to restore grasslands uses fences with many small paddocks and moving herds from one paddock to another after a day or two in order to mimic natural grazers and allowing the grass to grow optimally. It is estimated that increasing the carbon content of the soils in the world's 3.5 billion hectares of agricultural grassland would offset nearly 12 years of CO_2 emissions. Allan Savory, as part of holistic management, claims that while large herds are often blamed for desertification, prehistoric lands used to support large or larger herds and areas where herds were removed in the United States are still desertifying.

Land Degradation

Land degradation is a process in which the value of the biophysical environment is affected by a combination of human-induced processes acting upon the land. It is viewed as any change or disturbance to the land perceived to be deleterious or undesirable. Natural hazards are excluded as a cause; however human activities can indirectly affect phenomena such as floods and bush fires.

Serious land degradation in Nauru after the depletion of the phosphate cover through mining

This is considered to be an important topic of the 21st century due to the implications land degradation has upon agronomic productivity, the environment, and its effects on food security. It is estimated that up to 40% of the world's agricultural land is seriously degraded.

Measures

Land degradation is a broad term that can be applied differently across a wide range of scenarios. There are four main ways of looking at land degradation and its impact on the environment around it:

1. A temporary or permanent decline in the productive capacity of the land. This can be seen through a loss of biomass, a loss of actual productivity or in potential productivity, or a loss or change in vegetative cover and soil nutrients.

2. Action in the land's capacity to provide resources for human livelihoods. This can be measured from a base line of past land use.

3. Loss of biodiversity: A loss of range of species or ecosystem complexity as a decline in the environmental quality.

4. Shifting ecological risk: increased vulnerability of the environment or people to destruction or crisis. This is measured through a base line in the form of pre-existing risk of crisis or destruction.

A problem with defining land degradation is that what one group of people might view as degradation, others might view as a benefit or opportunity. For example, planting crops at a location with heavy rainfall and steep slopes would create scientific and environmental concern regarding the risk of soil erosion by water, yet farmers could view the location as a favourable one for high crop yields.

Different Types

Potato field with soil erosion

In addition to the usual types of land degradation that have been known for centuries (water, wind and mechanical erosion, physical, chemical and biological degradation), four other types have emerged in the last 50 years:

- pollution, often chemical, due to agricultural, industrial, mining or commercial activities;
- loss of arable land due to urban construction;
- artificial radioactivity, sometimes accidental;
- land-use constraints associated with armed conflicts.

Overall, 36 types of land degradation can be assessed. All are induced or aggravated by human activities, e.g. sheet erosion, silting, aridification, salinization, urbanization, etc.

Causes

Overgrazing by livestock can lead to land degradation

Land degradation is a global problem largely related to agricultural use. Causes include:

- Land clearance, such as clearcutting and deforestation
- Agricultural depletion of soil nutrients through poor farming practices
- Livestock including overgrazing and overdrafting
- Inappropriate irrigation and overdrafting

- Urban sprawl and commercial development

- Vehicle off-roading

- Quarrying of stone, sand, ore and minerals

- Increase in field size due to economies of scale, reducing shelter for wildlife, as hedgerows and copses disappear

- Exposure of naked soil after harvesting by heavy equipment

- Monoculture, destabilizing the local ecosystem

- Dumping of non-biodegradable trash, such as plastics

- Soil degradation, e.g.

  o   Soil contamination

  o   Soil erosion

  o   Soil acidification

  o   Loss of soil carbon

Effects

Soil erosion in a wheat field near Pullman, USA.

Overcutting of vegetation occurs when people cut forests, woodlands and shrublands—to obtain timber, fuelwood and other products—at a pace exceeding the rate of natural regrowth. This is frequent in semi-arid environments, where fuelwood shortages are often severe.

Overgrazing is the grazing of natural pastures at stocking intensities above the livestock carrying capacity; the resulting decrease in the vegetation cover is a leading cause of wind and water erosion. It is a significant factor in Afghanistan. The growing population pressure, during 1980-1990, has led to decreases in the already small areas of agricultural land per person in six out of eight countries (14% for India and 22% for Pakistan).

Population pressure also operates through other mechanisms. Improper agricultural practices, for instance, occur only under constraints such as the saturation of good lands under population pressure which leads settlers to cultivate too shallow or too steep soils, plough fallow land before it has recovered its fertility, or attempt to obtain multiple crops by irrigating unsuitable soils.

High population density is not always related to land degradation. Rather, it is the practices of the human population that can cause a landscape to become degraded. Populations can be a benefit to the land and make it more productive than it is in its natural state. Land degradation is an important factor of internal displacement in many African and Asian countries.

Severe land degradation affects a significant portion of the Earth's arable lands, decreasing the wealth and economic development of nations. As the land resource base becomes less productive, food security is compromised and competition for dwindling resources increases, the seeds of famine and potential conflict are sown.

Sensitivity and Resilience

Sensitivity and resilience are measures of the vulnerability of a landscape to degradation. These two factors combine to explain the degree of vulnerability. Sensitivity is the degree to which a land system undergoes change due to natural forces, human intervention or a combination of both. Resilience is the ability of a landscape to absorb change, without significantly altering the relationship between the relative importance and numbers of individuals and species that compose the community. It also refers to the ability of the region to return to its original state after being changed in some way. The resilience of a landscape can be increased or decreased through human interaction based upon different methods of land-use management. Land that is degraded becomes less resilient than undegraded land, which can lead to even further degradation through shocks to the landscape.

Climate Change

Significant land degradation from seawater inundation, particularly in river deltas and on low-lying islands, is a potential hazard that was identified in a 2007 IPCC report.

As a result of sea-level rise from climate change, salinity levels can reach levels where agriculture becomes impossible in very low-lying areas.

Habitat Destruction

Habitat destruction is the process in which natural habitat is rendered unable to support the species present. In this process, the organisms that previously used the site are displaced or destroyed, reducing biodiversity. Habitat destruction by human activity is mainly for the purpose of harvesting natural resources for industry production and urbanization. Clearing habitats for agriculture is the principal cause of habitat destruction. Other important causes of habitat destruction include mining, logging, trawling and urban sprawl. Habitat destruction is currently ranked as the primary cause of species extinction worldwide. It is a process of natural environmental change that may be caused by habitat fragmentation, geological processes, climate change or by human activities such as the introduction of invasive species, ecosystem nutrient depletion, and other human activities

The terms habitat loss and habitat reduction are also used in a wider sense, including loss of habitat from other factors, such as water and noise pollution.

Impacts on Organisms

In the simplest term, when a habitat is destroyed, the plants, animals, and other organisms that occupied the habitat have a reduced carrying capacity so that populations decline and extinction becomes more likely. Perhaps the greatest threat to organisms and biodiversity is the process of habitat loss. Temple (1986) found that 82% of endangered bird species were significantly threatened by habitat loss. Endemic organisms with limited ranges are most affected by habitat destruction, mainly because these organisms are not found anywhere else within the world and thus, have less chance of recovering. Many endemic organisms have very specific requirements for their survival that can only be found within a certain ecosystem, resulting in their extinction. Extinction may also take place very long after the destruction of habitat, a phenomenon known as extinction debt. Habitat destruction can also decrease the range of certain organism populations. This can result in the reduction of genetic diversity and perhaps the production of infertile youths, as these organisms would have a higher possibility of mating with related organisms within their population, or different species. One of the most famous examples is the impact upon China's giant panda, once found across the nation. Now it is only found in fragmented and isolated regions in the southwest of the country, as a result of widespread deforestation in the 20th century.

Geography

Satellite photograph of deforestation in Bolivia. Originally dry tropical forest,
the land is being cleared for soybean cultivation.

Biodiversity hotspots are chiefly tropical regions that feature high concentrations of endemic species and, when all hotspots are combined, may contain over half of the world's terrestrial species. These hotspots are suffering from habitat loss and destruction. Most of the natural habitat on islands and in areas of high human population density has already been destroyed (WRI, 2003). Islands suffering extreme habitat destruction include New Zealand, Madagascar, the Philippines, and Japan. South and East Asia — especially China, India, Malaysia, Indonesia, and Japan — and many areas in West Africa have extremely dense human populations that allow little room for natural habitat. Marine areas close to highly populated coastal cities also face degradation of their coral reefs or other marine habitat. These areas include the eastern coasts of Asia and Africa, northern coasts of South America, and the Caribbean Sea and its associated islands.

Regions of unsustainable agriculture or unstable governments, which may go hand-in-hand, typically experience high rates of habitat destruction. Central America, Sub-Saharan Africa, and the Amazonian tropical rainforest areas of South America are the main regions with unsustainable agricultural practices and/or government mismanagement.

Areas of high agricultural output tend to have the highest extent of habitat destruction. In the U.S., less than 25% of native vegetation remains in many parts of the East and Midwest. Only 15% of land area remains unmodified by human activities in all of Europe.

Ecosystems

Tropical rainforests have received most of the attention concerning the destruction of habitat. From the approximately 16 million square kilometers of tropical rainforest habitat that originally existed worldwide, less than 9 million square kilometers remain today. The current rate of deforestation is 160,000 square kilometers per year, which equates to a loss of approximately 1% of original forest habitat each year.

Jungle burned for agriculture in southern Mexico

Farmers near newly cleared land within Taman Nasional Kerinci Seblat (Kerinci Seblat National Park), Sumatra.

Other forest ecosystems have suffered as much or more destruction as tropical rainforests. Farming and logging have severely disturbed at least 94% of temperate broadleaf forests; many old growth forest stands have lost more than 98% of their previous area because of human activities. Tropical deciduous dry forests are easier to clear and burn and are more suitable for agriculture and cattle ranching than tropical rainforests; consequently, less than 0.1% of dry forests in Central America's Pacific Coast and less than 8% in Madagascar remain from their original extents.

Plains and desert areas have been degraded to a lesser extent. Only 10-20% of the world's drylands, which include temperate grasslands, savannas, and shrublands, scrub, and deciduous forests, have been somewhat degraded. But included in that 10-20% of land is the approximately 9 million square kilometers of seasonally dry-lands that humans have converted to deserts through the process of desertification. The tallgrass prairies of North America, on the other hand, have less than 3% of natural habitat remaining that has not been converted to farmland.

Wetlands and marine areas have endured high levels of habitat destruction. More than 50% of wetlands in the U.S. have been destroyed in just the last 200 years. Between 60% and 70% of European wetlands have been completely destroyed. About one-fifth (20%) of marine coastal areas have been highly modified by humans. One-fifth of coral reefs have also been destroyed, and another fifth has been severely degraded by overfishing, pollution, and invasive species; 90% of the Philippines' coral reefs alone have been destroyed. Finally, over 35% mangrove ecosystems worldwide have been destroyed.

Natural Causes

Habitat destruction through natural processes such as volcanism, fire, and climate change is well documented in the fossil record. One study shows that habitat fragmentation of tropical rainforests in Euramerica 300 million years ago led to a great loss of amphibian diversity, but simultaneously the drier climate spurred on a burst of diversity among reptiles.

Human Causes

Deforestation and roads in Amazonia, the Amazon Rainforest.

Habitat destruction caused by humans includes land conversion from forests, etc. to arable land, urban sprawl, infrastructure development, and other anthropogenic changes to the characteristics of land. Habitat degradation, fragmentation, and pollution are aspects of habitat destruction caused by humans that do not necessarily involve over destruction of habitat, yet result in habitat collapse. Desertification, deforestation, and coral reef degradation are specific types of habitat destruction for those areas (deserts, forests, coral reefs).

Geist and Lambin (2002) assessed 152 case studies of net losses of tropical forest cover to determine any patterns in the proximate and underlying causes of tropical deforestation. Their results, yielded as percentages of the case studies in which each parameter was a significant factor, provide a quantitative prioritization of which proximate and underlying causes were the most significant. The proximate causes were clustered into broad categories of agricultural expansion (96%), infrastructure expansion (72%), and wood extraction (67%). Therefore, according to this study, forest conversion to agriculture is the main land use change responsible for tropical deforestation. The specific categories reveal further insight into the specific causes of tropical deforestation: transport extension (64%), commercial wood extraction (52%), permanent cultivation (48%), cattle ranching (46%), shifting (slash and burn) cultivation (41%), subsistence agriculture (40%), and fuel wood extraction for domestic use (28%). One result is that shifting cultivation is not the primary cause of deforestation in all world regions, while transport extension (including the construction of new roads) is the largest single proximate factor responsible for deforestation.

Drivers

Nanjing Road in Shanghai

While the above-mentioned activities are the proximal or direct causes of habitat destruction in that they actually destroy habitat, this still does not identify why humans destroy habitat. The forces that cause humans to destroy habitat are known as *drivers* of habitat destruction. Demographic, economic, sociopolitical, scientific and technological, and cultural drivers all contribute to habitat destruction.

Demographic drivers include the expanding human population; rate of population increase over time; spatial distribution of people in a given area (urban versus rural), ecosystem type, and country; and the combined effects of poverty, age, family planning, gender, and education status of people in certain areas. Most of the exponential human population growth worldwide is occurring in or close to biodiversity hotspots. This may explain why human population density accounts for 87.9% of the variation in numbers of threatened species across 114 countries, providing indisputable evidence that people play the largest role in decreasing biodiversity. The boom in human population and migration of people into such species-rich regions are making conservation efforts not only more urgent but also more likely to conflict with local human interests. The high local population density in such areas is directly correlated to the poverty status of the local people, most of whom lacking an education and family planning.

From the Geist and Lambin (2002) study, the underlying driving forces were prioritized as follows (with the percent of the 152 cases the factor played a significant role in): economic factors (81%), institutional or policy factors (78%), technological factors (70%), cultural or socio-political factors (66%), and demographic factors (61%). The main economic factors included commercialization and growth of timber markets (68%), which are driven by national and international demands;

urban industrial growth (38%); low domestic costs for land, labor, fuel, and timber (32%); and increases in product prices mainly for cash crops (25%). Institutional and policy factors included formal pro-deforestation policies on land development (40%), economic growth including colonization and infrastructure improvement (34%), and subsidies for land-based activities (26%); property rights and land-tenure insecurity (44%); and policy failures such as corruption, lawlessness, or mismanagement (42%). The main technological factor was the poor application of technology in the wood industry (45%), which leads to wasteful logging practices. Within the broad category of cultural and sociopolitical factors are public attitudes and values (63%), individual/household behavior (53%), public unconcern toward forest environments (43%), missing basic values (36%), and unconcern by individuals (32%). Demographic factors were the in-migration of colonizing settlers into sparsely populated forest areas (38%) and growing population density — a result of the first factor — in those areas (25%).

There are also feedbacks and interactions among the proximate and underlying causes of deforestation that can amplify the process. Road construction has the largest feedback effect, because it interacts with—and leads to—the establishment of new settlements and more people, which causes a growth in wood (logging) and food markets. Growth in these markets, in turn, progresses the commercialization of agriculture and logging industries. When these industries become commercialized, they must become more efficient by utilizing larger or more modern machinery that often are worse on the habitat than traditional farming and logging methods. Either way, more land is cleared more rapidly for commercial markets. This common feedback example manifests just how closely related the proximate and underlying causes are to each other.

Impact on Human Population

The draining and development of coastal wetlands that previously protected the Gulf Coast contributed to severe flooding in New Orleans, Louisiana in the aftermath of Hurricane Katrina.

Habitat destruction vastly increases an area's vulnerability to natural disasters like flood and drought, crop failure, spread of disease, and water contamination. On the other hand, a healthy ecosystem with good management practices will reduce the chance of these events happening, or will at least mitigate adverse impacts.

Agricultural land can actually suffer from the destruction of the surrounding landscape. Over the past 50 years, the destruction of habitat surrounding agricultural land has degraded approximately 40% of agricultural land worldwide via erosion, salinization, compaction, nutrient depletion, pollution, and urbanization. Humans also lose direct uses of natural habitat when habitat is destroyed. Aesthetic uses such as birdwatching, recreational uses like hunting and fishing, and ecotourism usually rely upon virtually undisturbed habitat. Many people value the complexity of the natural world and are disturbed by the loss of natural habitats and animal or plant species worldwide.

Probably the most profound impact that habitat destruction has on people is the loss of many valuable ecosystem services. Habitat destruction has altered nitrogen, phosphorus, sulfur, and carbon cycles, which has increased the frequency and severity of acid rain, algal blooms, and fish kills in rivers and oceans and contributed tremendously to global climate change. One ecosystem service whose significance is becoming more realized is climate regulation. On a local scale, trees provide windbreaks and shade; on a regional scale, plant transpiration recycles rainwater and maintains constant annual rainfall; on a global scale, plants (especially trees from tropical rainforests) from around the world counter the accumulation of greenhouse gases in the atmosphere by sequestering carbon dioxide through photosynthesis. Other ecosystem services that are diminished or lost altogether as a result of habitat destruction include watershed management, nitrogen fixation, oxygen production, pollination, waste treatment (i.e., the breaking down and immobilization of toxic pollutants), and nutrient recycling of sewage or agricultural runoff.

The loss of trees from the tropical rainforests alone represents a substantial diminishing of the earth's ability to produce oxygen and use up carbon dioxide. These services are becoming even more important as increasing carbon dioxide levels is one of the main contributors to global climate change.

The loss of biodiversity may not directly affect humans, but the indirect effects of losing many species as well as the diversity of ecosystems in general are enormous. When biodiversity is lost, the environment loses many species that provide valuable and unique roles to the ecosystem. The environment and all its inhabitants rely on biodiversity to recover from extreme environmental conditions. When too much biodiversity is lost, a catastrophic event such as an earthquake, flood, or volcanic eruption could cause an ecosystem to crash, and humans would obviously suffer from that. Loss of biodiversity also means that humans are losing animals that could have served as biological control agents and plants that could potentially provide higher-yielding crop varieties, pharmaceutical drugs to cure existing or future diseases or cancer, and new resistant crop varieties for agricultural species susceptible to pesticide-resistant insects or virulent strains of fungi, viruses, and bacteria.

The negative effects of habitat destruction usually impact rural populations more directly than urban populations. Across the globe, poor people suffer the most when natural habitat is destroyed, because less natural habitat means less natural resources per capita, yet wealthier people and countries simply have to pay more to continue to receive more than their per capita share of natural resources.

Another way to view the negative effects of habitat destruction is to look at the opportunity cost of keeping an area undisturbed. In other words, what are people losing out on by taking away a given habitat? A country may increase its food supply by converting forest land to row-crop agriculture, but the value of the same land may be much larger when it can supply natural resources or services such as clean water, timber, ecotourism, or flood regulation and drought control.

Outlook

The rapid expansion of the global human population is increasing the world's food requirement substantially. Simple logic instructs that more people will require more food. In fact, as the world's population increases dramatically, agricultural output will need to increase by at least 50%, over the next 30 years. In the past, continually moving to new land and soils provided a boost in food production to appease the global food demand. That easy fix will no longer be available, however, as more than 98% of all land suitable for agriculture is already in use or degraded beyond repair.

The impending global food crisis will be a major source of habitat destruction. Commercial farmers are going to become desperate to produce more food from the same amount of land, so they will use more fertilizers and less concern for the environment to meet the market demand. Others will seek out new land or will convert other land-uses to agriculture. Agricultural intensification will become widespread at the cost of the environment and its inhabitants. Species will be pushed out of their habitat either directly by habitat destruction or indirectly by fragmentation, degradation, or pollution. Any efforts to protect the world's remaining natural habitat and biodiversity will compete directly with humans' growing demand for natural resources, especially new agricultural lands.

Solutions

Chelonia mydas on a Hawaiian coral reef. Although the endangered species is protected, habitat loss from human development is a major reason for the loss of green turtle nesting beaches.

In most cases of tropical deforestation, three to four underlying causes are driving two to three proximate causes. This means that a universal policy for controlling tropical deforestation would not be able to address the unique combination of proximate and underlying causes of deforestation in each country. Before any local, national, or international deforestation policies are written and enforced, governmental leaders must acquire a detailed understanding of the complex combination of proximate causes and underlying driving forces of deforestation in a given area or country. This concept, along with many other results about tropical deforestation from the Geist and Lambin study, can easily be applied to habitat destruction in general. Governmental leaders need to take action by addressing the underlying driving forces, rather than merely regulating the proximate causes. In a broader sense, governmental bodies at a local, national, and international scale need to emphasize the following:

1. Considering the many irreplaceable ecosystem services provided by natural habitats.

2. Protecting remaining intact sections of natural habitat.

3. Educating the public about the importance of natural habitat and biodiversity.

4. Developing family planning programs in areas of rapid population growth.

5. Finding ecological ways to increase agricultural output without increasing the total land in production.

6. Preserving habitat corridors to minimize prior damage from fragmented habitats.

7. Reduce human population and expansion.

Endangered Species

The Iberian lynx (*Lynx pardinus*), an endangered species.

An endangered species is a species which has been categorized as likely to become extinct. Endangered (EN), as categorized by the International Union for Conservation of Nature (IUCN) Red List, is the second most severe conservation status for wild populations in the IUCN's schema after Critically Endangered (CR).

In 2012, the IUCN Red List featured 3079 animal and 2655 plant species as endangered (EN) worldwide. The figures for 1998 were, respectively, 1102 and 1197.

Many nations have laws that protect conservation-reliant species: for example, forbidding hunting, restricting land development or creating preserves. Population numbers, trends and species' conservation status can be found in the lists of organisms by population.

Conservation Status

The conservation status of a species indicates the likelihood that it will become extinct. Many factors are considered when assessing the conservation status of a species; e.g., such statistics as the number remaining, the overall increase or decrease in the population over time, breeding success rates, or known threats. The IUCN Red List of Threatened Species is the best-known worldwide conservation status listing and ranking system.

Over 40% of the world's species are estimated to be at risk of extinction. Internationally, 199 countries have signed an accord to create Biodiversity Action Plans that will protect endangered and other threatened species. In the United States, such plans are usually called Species Recovery Plans.

IUCN Red List

Though labelled a list, the IUCN Red List is a system of assessing the global conservation status of species that includes "Data Deficient" (DD) species – species for which more data and assessment is required before their status may be determined – as well species comprehensively assessed by the IUCN's species assessment process. Those species of "Near Threatened" (NT) and "Least Concern" (LC) status have been assessed and found to have relatively robust and healthy populations, though these may be in decline. Unlike their more general use elsewhere, the List uses the terms "endangered species" and "threatened species" with particular meanings: "Endangered" (EN) species lie between "Vulnerable" (VU) and "Critically Endangered" (CR) species, while "Threatened" species are those species determined to be Vulnerable, Endangered or Critically Endangered.

Kemp's ridley sea turtle, an endangered species

The IUCN categories, with examples of animals classified by them, include:

Extinct (EX)

Examples: aurochs, Bali tiger, blackfin cisco, Caribbean monk seal, Carolina parakeet Caspian tiger, dodo, dusky seaside sparrow, eastern cougar, golden toad, great auk, Japanese sea lion, Javan tiger, Labrador duck, passenger pigeon, Schomburgk's deer, Steller's sea cow, thylacine, toolache wallaby, western black rhinoceros, California Grizzly Bear

Extinct in the wild (EW)

Captive individuals survive, but there is no free-living, natural population.

Examples: Barbary lion, Hawaiian crow, Père David's deer, scimitar oryx, Socorro dove, Wyoming toad

Critically endangered (CR)

Brown spider monkey, an endangered species

Faces an extremely high risk of extinction in the immediate future.

Examples: addax, African wild ass, Alabama cavefish, Amur leopard, Arabian leopard, Arakan forest turtle, Asiatic cheetah, axolotl, Bactrian camel, black rhino, blue-throated macaw, Brazilian merganser, brown spider monkey, California condor, Chinese alligator, Chinese giant salamander, Cross River gorilla, Florida panther, gharial, Hawaiian monk seal, Imperial woodpecker, Ivory-billed Woodpecker, Javan rhino, kakapo, Leadbeater's possum, Mediterranean monk seal, mountain gorilla, Northwest African cheetah, northern hairy-nosed wombat, Philippine eagle, red wolf, saiga, Siamese crocodile, Malayan tiger, Spix's macaw, southern bluefin tuna, South China tiger, Sumatran elephant, Sumatran orangutan, Sumatran rhinoceros, Sumatran tiger, vaquita, Yangtze river dolphin, northern white rhinoceros

Endangered (EN)

The Siberian tiger is an Endangered (EN) tiger subspecies. Three tiger subspecies are already extinct

Faces a high risk of extinction in the near future.

Examples: African penguin, African wild dog, Amur tiger, Asian elephant, Asiatic lion, Australasian bittern, blue whale, bonobo, Bornean orangutan, common chimpanzee, dhole,eastern low land gorilla, Ethiopian wolf, Flores crow, hispid hare, giant otter, Goliath frog, green sea turtle, loggerhead sea turtle, Grevy's zebra, hyacinth macaw, Humblot's heron, Iberian lynx, Japanese crane, Japanese night heron, Lear's macaw, Malayan tapir, markhor, Malagasy pond heron, Persian leopard, proboscis monkey, purple-faced langur, pygmy hippopotamus, red-breasted goose, Rothschild's giraffe, snow leopard, South Andean deer, Sri Lankan elephant, takhi, Toque macaque, Vietnamese pheasant, volcano rabbit, wild water buffalo, white-eared night heron, fishing cat, tasmanian devil, red panda, whale shark

Vulnerable (VU)

Siamese crocodile, an endangered species

Faces a high risk of endangerment in the medium term.

Examples: African grey parrot, African bush elephant, African leopard, African lion, American paddlefish, common carp, clouded leopard, cheetah, dugong, Far Eastern curlew, fossa, Galapagos tortois, gaur, giraffe, blue-eyed cockatoo, golden hamster, Great slaty woodpecker, hippopotamus, Humboldt penguin, Indian rhinoceros, Komodo dragon, lesser white-fronted goose, mandrill, maned sloth, mountain zebra, Hawaiian goose, polar bear, sloth bear, takin, yak, great white shark, American crocodile, white-necked crow, dingo, king cobra

Near-threatened (NT)

May be considered threatened in the near future.

Examples: American bison, Asian golden cat, blue-billed duck, emperor goose, emperor penguin, Eurasian curlew, jaguar, Larch Mountain salamander, Magellanic penguin, maned wolf, narwhal, margay, montane solitary eagle, Pampas cat, Pallas's cat, reddish egret, white rhinoceros, striped hyena, tiger shark, white eared pheasant

Least concern (LC)

Blue-throated macaw, an endangered species

No immediate threat to species' survival.

Examples: American alligator, American crow, Indian peafowl, olive baboon, bald eagle, brown bear, brown rat, brown-throated sloth, Canada goose, cane toad, common wood pigeon, cougar, common frog, grey wolf, house mouse, wolverine, palm cockatoo, mallard, meerkat, mute swan, platypus, red-billed quelea, red-tailed hawk, rock pigeon, scarlet macaw, southern elephant seal, milk shark, red howler monkey

Criteria for 'Endangered (EN)'

A) Reduction in population size based on any of the following:

1. An observed, estimated, inferred or suspected population size reduction of ≥ 70% over the last 10 years or three generations, whichever is the longer, where the causes of the reduction are clearly reversible AND understood AND ceased, based on (and specifying) any of the following:

    1. direct observation

    2. an index of abundance appropriate for the taxon

    3. a decline in area of occupancy, extent of occurrence or quality of habitat

    4. actual or potential levels of exploitation

    5. the effects of introduced taxa, hybridisation, pathogens, pollutants, competitors or parasites.

2. An observed, estimated, inferred or suspected population size reduction of ≥ 50% over the last 10 years or three generations, whichever is the longer, where the reduction or its causes may not have ceased OR may not be understood OR may not be reversible, based on (and specifying) any of (a) to (e) under A1.

3. A population size reduction of ≥ 50%, projected or suspected to be met within the next 10 years or three generations, whichever is the longer (up to a maximum of 100 years), based on (and specifying) any of (b) to (e) under A1.

4. An observed, estimated, inferred, projected or suspected population size reduction of ≥ 50% over any 10 year or three generation period, whichever is longer (up to a maximum of 100 years in the future), where the time period must include both the past and the future, and where the reduction or its causes may not have ceased OR may not be understood OR may not be reversible, based on (and specifying) any of (a) to (e) under A1.

B) Geographic range in the form of either B1 (extent of occurrence) OR B2 (area of occupancy) OR both:

1. Extent of occurrence estimated to be less than 5,000 km², and estimates indicating at least two of a-c;

    1. Severely fragmented or known to exist at no more than five locations.

    2. Continuing decline, inferred, observed or projected, in any of the following:

        1. extent of occurrence

    2. area of occupancy

    3. area, extent or quality of habitat

    4. number of locations or subpopulations

    5. number of mature individuals

3. Extreme fluctuations in any of the following:

    1. extent of occurrence

    2. area of occupancy

    3. number of locations or subpopulations

    4. number of mature individuals

2. Area of occupancy estimated to be less than 500 km^2, and estimates indicating at least two of a-c

    1. Severely fragmented or known to exist at no more than five locations.

    2. Continuing decline, inferred, observed or projected, in any of the following:

        1. extent of occurrence

        2. area of occupancy

        3. area, extent or quality of habitat

        4. number of locations or subpopulations

        5. number of mature individuals

    3. Extreme fluctuations in any of the following:

        1. extent of occurrence

        2. area of occupancy

        3. number of locations or subpopulations

        4. number of mature individuals

C) Population estimated to number fewer than 2,500 mature individuals and either:

1. An estimated continuing decline of at least 20% within five years or two generations, whichever is longer, (up to a maximum of 100 years in the future) OR

2. A continuing decline, observed, projected, or inferred, in numbers of mature individuals AND at least one of the follow (a-b):

- Population structure in the form of one of the following:

- no subpopulation estimated to contain more than 250 mature individuals, OR

- at least 95% of mature individuals in one subpopulation

- Extreme fluctuations in number of mature individuals

D) Population size estimated to number fewer than 250 mature individuals.

E) Quantitative analysis showing the probability of extinction in the wild is at least 20% within 20 years or five generations, whichever is the longer (up to a maximum of 100 years).

Endangered Species in the United States

There is data from the United States that shows a correlation between human populations and threatened and endangered species. Using species data from the Database on the Economics and Management of Endangered Specwies (DEMES) database and the period that the Endangered Species Act (ESA) has been in existence, 1970 to 1997, a table was created that suggests a positive relationship between human activity and species endangerment.

Endangered Species Act

"Endangered" in relation to "threatened" under the ESA.

Under the Endangered Species Act in the United States, species may be listed as "endangered" or "threatened". The Salt Creek tiger beetle (*Cicindela nevadica lincolniana*) is an example of an endangered subspecies protected under the ESA. The US Fish and Wildlife Service as well as the National Marine Fisheries Service are held responsible for classifying and protecting endangered species, and adding a particular species to the list can be a long, controversial process.

Some endangered species laws are controversial. Typical areas of controversy include: criteria for placing a species on the endangered species list and criteria for removing a species from the list once its population has recovered; whether restrictions on land development constitute a "taking" of land by the government; the related question of whether private landowners should be compensated for the loss of uses of their lands; and obtaining reasonable exceptions to protection laws. Also lobbying from hunters and various industries like the petroleum industry, construction industry, and logging, has been an obstacle in establishing endangered species laws.

The Bush administration lifted a policy that required federal officials to consult a wildlife expert before taking actions that could damage endangered species. Under the Obama administration, this policy has been reinstated.

Being listed as an endangered species can have negative effect since it could make a species more desirable for collectors and poachers. This effect is potentially reducible, such as in China where commercially farmed turtles may be reducing some of the pressure to poach endangered species.

Another problem with the listing species is its effect of inciting the use of the "shoot, shovel, and shut-up" method of clearing endangered species from an area of land. Some landowners currently may perceive a diminution in value for their land after finding an endangered animal on it. They have allegedly opted to silently kill and bury the animals or destroy habitat, thus removing the problem from their land, but at the same time further reducing the population of an endangered species. The effectiveness of the Endangered Species Act – which coined the term «endangered species» – has been questioned by business advocacy groups and their publications but is nevertheless widely recognized by wildlife scientists who work with the species as an effective recovery tool. Nineteen species have been delisted and recovered and 93% of listed species in the northeastern United States have a recovering or stable population.

Currently, 1,556 known species in the world have been identified as near extinction or endangered and are under protection by government law. This approximation, however, does not take into consideration the number of species threatened with endangerment that are not included under the protection of such laws as the Endangered Species Act. According to NatureServe's global conservation status, approximately thirteen percent of vertebrates (excluding marine fish), seventeen percent of vascular plants, and six to eighteen percent of fungi are considered imperiled. Thus, in total, between seven and eighteen percent of the United States' known animals, fungi and plants are near extinction. This total is substantially more than the number of species protected in the United States under the Endangered Species Act.

Bald eagle

American bison

Ever since mankind began hunting to preserve itself, over-hunting and fishing has been a large and dangerous problem. Of all the species who became extinct due to interference from mankind, the dodo, passenger pigeon, great auk, Tasmanian tiger and Steller's sea cow are some of the more well known examples; with the bald eagle, grizzly bear, American bison, Eastern timber wolf and sea turtle having been hunted to near-extinction. Many began as food sources seen as necessary for survival but became the target of sport. However, due to major efforts to prevent extinction, the bald eagle, or *Haliaeetus leucocephalus* is now under the category of Least Concern on the red list. A present-day example of the over-hunting of a species can be seen in the oceans as populations of certain whales have been greatly reduced. Large whales like the blue whale, bowhead whale, finback whale, gray whale, sperm whale and humpback whale are some of the eight whales which are currently still included on the Endangered Species List. Actions have been taken to attempt reduction in whaling and increase population sizes, including prohibiting all whaling in United States waters, the formation of the CITES treaty which protects all whales, along with the formation of the International Whaling Commission (IWC). But even though all of these movements have been put in place, countries such as Japan continue to hunt and harvest whales under the claim of "scientific purposes". Over-hunting, climatic change and habitat loss leads in landing species in endangered species list and could mean that extinction rates could increase to a large extent in the future.

Invasive Species

The introduction of non-indigenous species to an area can disrupt the ecosystem to such an extent that native species become endangered. Such introductions may be termed alien or invasive species. In some cases the invasive species compete with the native species for food or prey on the natives. In other cases a stable ecological balance may be upset by predation or other causes leading to unexpected species decline. New species may also carry diseases to which the native species have no resistance.

Conservation

The dhole, Asia's most endangered top predator, is on the edge of extinction.

Captive Breeding

Captive breeding is the process of breeding rare or endangered species in human controlled environments with restricted settings, such as wildlife reserves, zoos and other conservation facilities. Captive breeding is meant to save species from extinction and so stabilize the population of the species that it will not disappear.

This technique has worked for many species for some time, with probably the oldest known such instances of captive mating being attributed to menageries of European and Asian rulers, an example being the Père David's deer. However, captive breeding techniques are usually difficult to implement for such highly mobile species as some migratory birds (e.g. cranes) and fishes (e.g. hilsa). Additionally, if the captive breeding population is too small, then inbreeding may occur due to a reduced gene pool and reduce immunity.

In 1981, the Association of Zoos and Aquariums (AZA) created a Species Survival Plan (SSP) in order to help preserve specific endangered and threatened species through captive breeding. With over 450 SSP Plans, there are a number of endangered species that are covered by the AZA with plans to cover population management goals and recommendations for breeding for a diverse and healthy population, created by Taxon Advisory Groups. These programs are commonly created as a last resort effort. SSP Programs regularly participate in species recovery, veterinary care for wildlife disease outbreaks, and a number of other wildlife conservation efforts. The AZA's Species Survival Plan also has breeding and transfer programs, both within and outside of AZA - certified zoos and aquariums. Some animals that are part of SSP programs are giant pandas, lowland gorillas, and California condors.

Private Farming

Black rhino

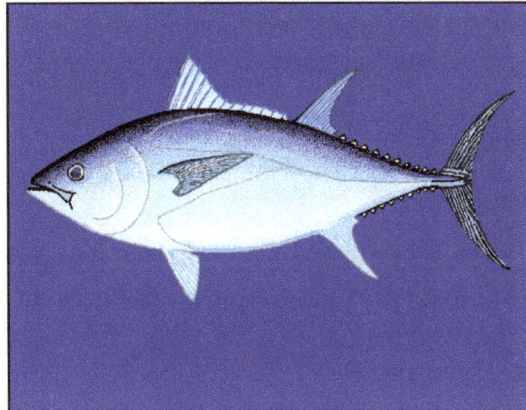

Southern bluefin tuna

Whereas poaching substantially reduces endangered animal populations, legal, for-profit, private farming does the opposite. It has substantially increased the populations of the southern black rhinoceros and southern white rhinoceros. Dr Richard Emslie, a scientific officer at the IUCN, said of such programs, "Effective law enforcement has become much easier now that the animals are largely privately owned... We have been able to bring local communities into the conservation programmes. There are increasingly strong economic incentives attached to looking after rhinos rather than simply poaching: from Eco-tourism or selling them on for a profit. So many owners are keeping them secure. The private sector has been key to helping our work."

Conservation experts view the effect of China's turtle farming on the wild turtle populations of China and South-Eastern Asia – many of which are endangered – as "poorly understood". Although they commend the gradual replacement of turtles caught wild with farm-raised turtles in the marketplace – the percentage of farm-raised individuals in the "visible" trade grew from around 30% in 2000 to around 70% in 2007 – they worry that many wild animals are caught to provide farmers with breeding stock. The conservation expert Peter Paul van Dijk noted that turtle farmers often believe that animals caught wild are superior breeding stock. Turtle farmers may, therefore, seek and catch the last remaining wild specimens of some endangered turtle species.

In 2009, researchers in Australia managed to coax southern bluefin tuna to breed in landlocked tanks, raising the possibility that fish farming may be able to save the species from overfishing.

Deforestation and Climate Change

Deforestation is one of the main contributors to climate change. It is the second largest anthropogenic source of carbon dioxide to the atmosphere, after fossil fuel combustion. Deforestation and forest degradation contribute to atmospheric greenhouse gas emissions through combustion of forest biomass and decomposition of remaining plant material and soil carbon. It used to account for more than 20% of carbon dioxide emissions, but it's currently somewhere around the 10% mark. By 2008, deforestation was 12% of total CO_2, or 15% if peatlands are included. These proportions are likely to have fallen since given the continued rise of fossil fuel use.

Averaged over all land and ocean surfaces, temperatures warmed roughly 1.53 °F (0.85 °C) between 1880 and 2012, according to the Intergovernmental Panel on Climate Change. In the Northern Hemisphere, 1983 to 2012 were the warmest 30-year period of the last 1400 years.

Effect on Climate Change

Decrease in Biodiversity

A 2007 study conducted by the National Science Foundation found that biodiversity and genetic diversity are codependent—that diversity among species requires diversity within a species, and vice versa. "If any one type is removed from the system, the cycle can break down, and the community becomes dominated by a single species."

Counteracting Climate Change

Reforestation

Reforestation is the natural or intentional restocking of existing forests and woodlands that have been depleted, usually through deforestation. It is the reestablishment of forest cover either naturally or artificially. Similar to the other methods of forestation, reforestation can be very effective because a single tree can absorb as much as 22 kilograms (48 lb) of carbon dioxide per year and can sequester 0.91 tonnes (1 short ton) of carbon dioxide by the time it reaches 40 years old.

Afforestation

Afforestation is the establishment of a forest or stand of trees in an area where there was no forest.

China

Although China has set official goals for reforestation, these goals were set for an 80-year time horizon and were not significantly met by 2008. China is trying to correct these problems with projects such as the Green Wall of China, which aims to replant forests and halt the expansion of the Gobi Desert. A law promulgated in 1981 requires that every school student over the age of 11 plant at least one tree per year. But average success rates, especially in state-sponsored plantings, remains relatively low. And even the properly planted trees have had great difficulty surviving the combined impacts of prolonged droughts, pest infestation and fires. Nonetheless, China currently has the highest afforestation rate of any country or region in the world, with 4.77 million hectares (47,000 square kilometers) of afforestation in 2008.

Japan

The primary goal of afforestation projects in Japan is to develop the forest structure of the nation and to maintain the biodiversity found in the Japanese wilderness. The Japanese temperate rainforest is scattered throughout the Japanese archipelago and is home to many endemic species that are not naturally found anywhere else. As development of the country's caused a decline in forest cover, a reduction in biodiversity was seen in those areas.

Agroforestry

Agroforestry or agro-sylviculture is a land use management system in which trees or shrubs are grown around or among crops or pastureland. It combines agricultural and forestry technologies to create more diverse, productive, profitable, healthy, and sustainable land-use systems.

Projects and Foundations

Arbor Day Foundation

Founded in 1972, the centennial of the first Arbor Day observance in the 19th century, the Foundation has grown to become the largest nonprofit membership organization dedicated to planting trees, with over one million members, supporters, and valued partners. They work on projects focused on planting trees around campuses, low-income communities, and communities that have been affected by natural disasters among other places.

Billion Tree Campaign

The Billion Tree Campaign was launched in 2006 by the United Nations Environment Programme (UNEP) as a response to the challenges of global warming, as well as to a wider array of sustainability challenges, from water supply to biodiversity loss. Its initial target was the planting of one billion trees in 2007. Only one year later in 2008, the campaign's objective was raised to 7 billion trees—a target to be met by the climate change conference that was held in Copenhagen, Denmark in December 2009. Three months before the conference, the 7 billion planted trees mark had been surpassed. In December 2011, after more than 12 billion trees had been planted, UNEP formally handed management of the program over to the not-for-profit Plant-for-the-Planet initiative, based in Munich, Germany.

The Amazon Fund (Brazil)

Considered the largest reserve of biological diversity in the world, the Amazon Basin is also the largest

Brazilian biome, taking up almost half the nation's territory. The Amazon Basin corresponds to two fifths of South America's territory. Its area of approximately seven million square kilometers covers the largest hydrographic network on the planet, through which runs about one fifth of the fresh water on the world's surface. Deforestation in the Amazon rainforest is a major cause to climate change due to the decreasing number of trees available to capture increasing carbon dioxide levels in the atmosphere.

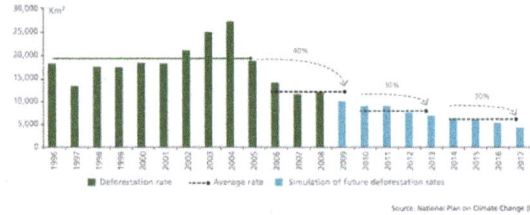

Four-year plan to reduce in deforestation in the Amazon

The Amazon Fund is aimed at raising donations for non-reimbursable investments in efforts to prevent, monitor and combat deforestation, as well as to promote the preservation and sustainable use of forests in the Amazon Biome, under the terms of Decree N.º 6,527, dated August 1, 2008. The Amazon Fund supports the following areas: management of public forests and protected areas, environmental control, monitoring and inspection, sustainable forest management, economic activities created with sustainable use of forests, ecological and economic zoning, territorial arrangement and agricultural regulation, preservation and sustainable use of biodiversity, and recovery of deforested areas. Besides those, the Amazon Fund may use up to 20% of its donations to support the development of systems to monitor and control deforestation in other Brazilian biomes and in biomes of other tropical countries.

Global Warming

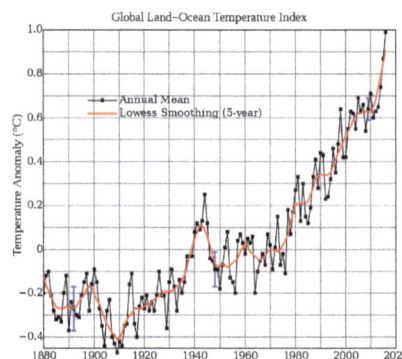

Global mean surface-temperature change from 1880 to 2016, relative to the 1951–1980 mean. The black line is the global annual mean and the red line is the five-year lowess smooth.
The blue uncertainty bars show a 95% confidence interval.

Global warming, also referred to as climate change, is the observed century-scale rise in the average temperature of the Earth's climate system and its related effects. Multiple lines of scientific evidence show that the climate system is warming. Many of the observed changes since the 1950s are unprecedented in the instrumental temperature record which extends back to the mid-19th century, and in paleoclimate proxy records over thousands of years.

In 2013, the Intergovernmental Panel on Climate Change (IPCC) Fifth Assessment Report concluded that "It is *extremely likely* that human influence has been the dominant cause of the observed warming since the mid-20th century." The largest human influence has been emission of greenhouse gases such as carbon dioxide, methane and nitrous oxide. Climate model projections summarized in the report indicated that during the 21st century the global surface temperature is likely to rise a further 0.3 to 1.7 °C (0.5 to 3.1 °F) for their lowest emissions scenario and 2.6 to 4.8 °C (4.7 to 8.6 °F) for the highest emissions scenario. These findings have been recognized by the national science academies of the major industrialized nations and are not disputed by any scientific body of national or international standing.

Future climate change and associated impacts will differ from region to region around the globe. Anticipated effects include warming global temperature, rising sea levels, changing precipitation, and expansion of deserts in the subtropics. Warming is expected to be greater over land than over the oceans and greatest in the Arctic, with the continuing retreat of glaciers, permafrost and sea ice. Other likely changes include more frequent extreme weather events including heat waves, droughts, heavy rainfall with floods and heavy snowfall; ocean acidification; and species extinctions due to shifting temperature regimes. Effects significant to humans include the threat to food security from decreasing crop yields and the abandonment of populated areas due to rising sea levels. Because the climate system has a large "inertia" and greenhouse gases will stay in the atmosphere for a long time, many of these effects will not only exist for decades or centuries, but will persist for tens of thousands of years.

Possible societal responses to global warming include mitigation by emissions reduction, adaptation to its effects, building systems resilient to its effects, and possible future climate engineering. Most countries are parties to the United Nations Framework Convention on Climate Change (UNFCCC), whose ultimate objective is to prevent dangerous anthropogenic climate change. Parties to the UNFCCC have agreed that deep cuts in emissions are required and that global warming should be limited to well below 2.0 °C (3.6 °F) relative to pre-industrial levels, with efforts made to limit warming to 1.5 °C (2.7 °F).

Public reactions to global warming and concern about its effects are also increasing. A global 2015 Pew Research Center report showed a median of 54% consider it "a very serious problem". There are significant regional differences, with Americans and Chinese (whose economies are responsible for the greatest annual CO_2 emissions) among the least concerned.

Observed Temperature Changes

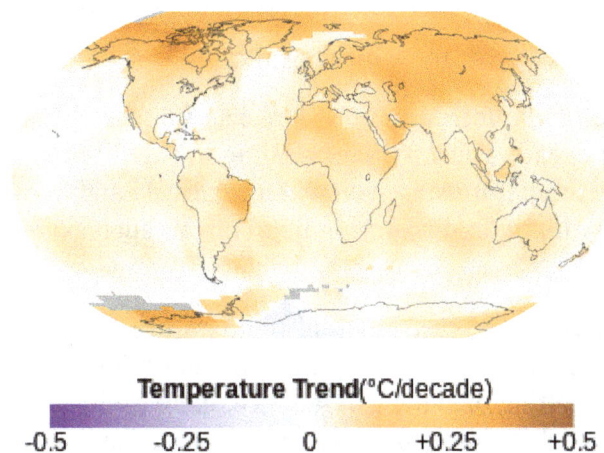

World map showing surface temperature trends (°C per decade) between 1950 and 2014.

The global average (land and ocean) surface temperature shows a warming of 0.85 [0.65 to 1.06] °C in the period 1880 to 2012, based on multiple independently produced datasets. Earth's average surface temperature rose by 0.74±0.18 °C over the period 1906–2005. The rate of warming almost doubled for the last half of that period (0.13±0.03 °C per decade, versus 0.07±0.02 °C per decade). Although the increase of near-surface atmospheric temperature is the measure of global warming often reported in the popular press, most of the additional energy stored in the climate system since 1970 has gone into the oceans. The rest has melted ice and warmed the continents and atmosphere.

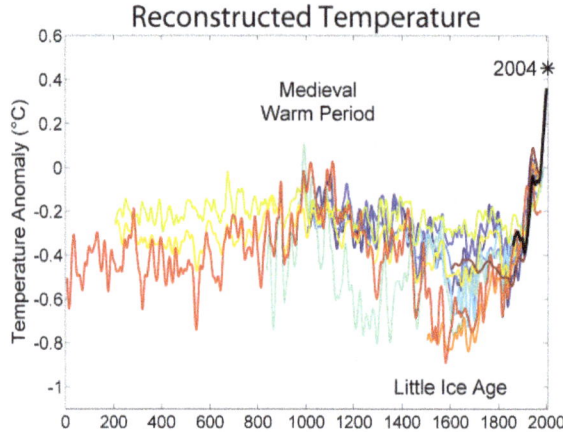

Two millennia of mean surface temperatures according to different reconstructions from climate proxies, each smoothed on a decadal scale, with the instrumental temperature record overlaid in black.

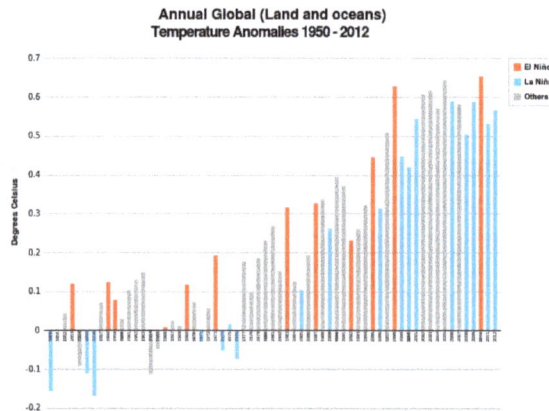

NOAA graph of global annual temperature anomalies 1950–2012, showing the El Niño Southern Oscillation

The average temperature of the lower troposphere has increased between 0.12 and 0.135 °C (0.216 and 0.243 °F) per decade since 1979, according to satellite temperature measurements. Climate proxies show the temperature to have been relatively stable over the one or two thousand years before 1850, with regionally varying fluctuations such as the Medieval Warm Period and the Little Ice Age.

The warming that is evident in the instrumental temperature record is consistent with a wide range of observations, as documented by many independent scientific groups. Examples include sea level rise, widespread melting of snow and land ice, increased heat content of the oceans, increased humidity, and the earlier timing of spring events, e.g., the flowering of plants. The probability that these changes could have occurred by chance is virtually zero.

Trends

Temperature changes vary over the globe. Since 1979, land temperatures have increased about twice as fast as ocean temperatures (0.25 °C per decade against 0.13 °C per decade). Ocean temperatures increase more slowly than land temperatures because of the larger effective heat capacity of the oceans and because the ocean loses more heat by evaporation. Since the beginning of industrialisation the temperature difference between the hemispheres has increased due to melting of sea ice and snow in the North. Average arctic temperatures have been increasing at almost twice the rate of the rest of the world in the past 100 years; however arctic temperatures are also highly variable. Although more greenhouse gases are emitted in the Northern than Southern Hemisphere this does not contribute to the difference in warming because the major greenhouse gases persist long enough to mix between hemispheres.

The thermal inertia of the oceans and slow responses of other indirect effects mean that climate can take centuries or longer to adjust to changes in forcing. One climate commitment study concluded that if greenhouse gases were stabilized at year 2000 levels, surface temperatures would still increase by about one-half degree Celsius, and another found that if they were stabilized at 2005 levels surface warming could exceed a whole degree Celsius. Some of this surface warming will be driven by past natural forcings which are still seeking equilibrium in the climate system. One study using a highly simplified climate model indicates these past natural forcings may account for as much as 64% of the committed 2050 surface warming and their influence will fade with time compared to the human contribution.

Global temperature is subject to short-term fluctuations that overlay long-term trends and can temporarily mask them. The relative stability in surface temperature from 2002 to 2009, which has been dubbed the global warming hiatus by the media and some scientists, is consistent with such an episode. 2015 updates to account for differing methods of measuring ocean surface temperature measurements show a positive trend over the recent decade.

Warmest Years

Sixteen of the 17 warmest years on record have occurred since 2000. While record-breaking years attract considerable public interest, individual years are less significant than the overall trend. Some climatologists have criticized the attention that the popular press gives to "warmest year" statistics. In particular, ocean oscillations such as the El Niño Southern Oscillation (ENSO) can cause temperatures of a given year to be abnormally warm or cold for reasons unrelated to the overall trend of climate change. Gavin Schmidt stated "the long-term trends or the expected sequence of records are far more important than whether any single year is a record or not."

Initial Causes of Temperature Changes (external forcings)

The climate system can spontaneously generate changes in global temperature for years to decades at a time but long-term changes in global temperature require *external forcings*. These forcings are "external" to the climate system but not necessarily external to Earth. Examples of external forcings include changes in atmospheric composition (e.g., increased concentrations of greenhouse gases), solar luminosity, volcanic eruptions, and variations in Earth's orbit around the Sun.

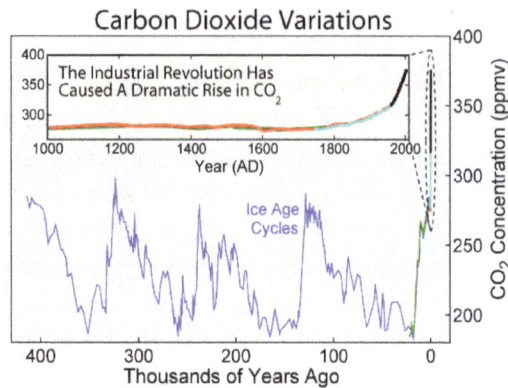

Carbon Dioxide Variations

CO_2 concentrations over the last 400,000 years.

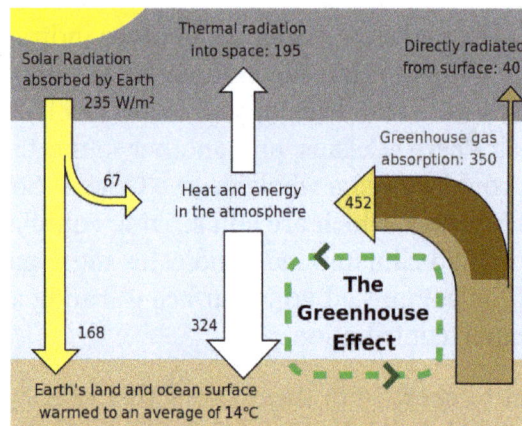

Greenhouse effect schematic showing energy flows between space,
the atmosphere, and Earth's surface. Energy exchanges are expressed
in watts per square metre (W/m²).

Greenhouse Gases

The greenhouse effect is the process by which absorption and emission of infrared radiation by gases in a planet's atmosphere warm its lower atmosphere and surface. It was proposed by Joseph Fourier in 1824, discovered in 1860 by John Tyndall, was first investigated quantitatively by Svante Arrhenius in 1896, and its scientific description was developed in the 1930s through 1960s by Guy Stewart Callendar.

On Earth, an atmosphere containing naturally occurring amounts of greenhouse gases causes air temperature near the surface to be about 33 °C (59 °F) warmer than it would be in their absence. Without the Earth's atmosphere, the Earth's average temperature would be well below the freezing temperature of water. The major greenhouse gases are water vapour, which causes about 36–70% of the greenhouse effect; carbon dioxide (CO_2), which causes 9–26%; methane (CH_4), which causes 4–9%; and ozone (O_3), which causes 3–7%. Clouds also affect the radiation balance through cloud forcings similar to greenhouse gases.

Human activity since the Industrial Revolution has increased the amount of greenhouse gases in the atmosphere, leading to increased radiative forcing from CO_2, methane, tropospheric ozone, CFCs and nitrous oxide. According to work published in 2007, the concentrations of CO_2 and methane had increased by 36% and 148% respectively since 1750. These levels are much higher than at any time during the last 800,000 years, the period for which reliable data has been extracted from ice cores. Less direct geological evidence indicates that CO_2 values higher than this were last seen about 20 million years ago.

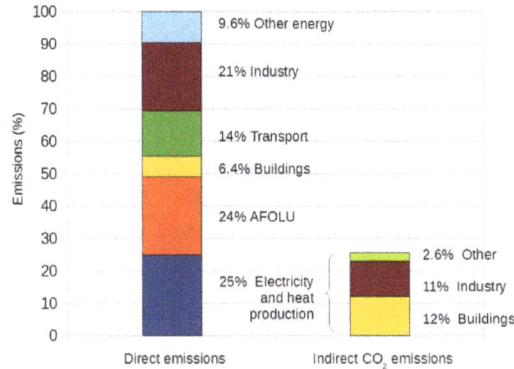

Annual world greenhouse gas emissions, in 2010, by sector.

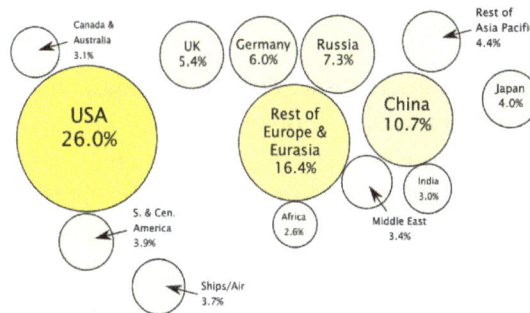

Percentage share of global cumulative energy-related CO_2 emissions between 1751 and 2012 across different regions.

Fossil fuel burning has produced about three-quarters of the increase in CO_2 from human activity over the past 20 years. The rest of this increase is caused mostly by changes in land-use, particularly deforestation. Another significant non-fuel source of anthropogenic CO_2 emissions is the calcination of limestone for clinker production, a chemical process which releases CO_2. Estimates of global CO_2 emissions in 2011 from fossil fuel combustion, including cement production and gas flaring, was 34.8 billion tonnes (9.5 ± 0.5 PgC), an increase of 54% above emissions in 1990. Coal burning was responsible for 43% of the total emissions, oil 34%, gas 18%, cement 4.9% and gas flaring 0.7%

In May 2013, it was reported that readings for CO_2 taken at the world's primary benchmark site in Mauna Loa surpassed 400 ppm. According to professor Brian Hoskins, this is likely the first time CO_2 levels have been this high for about 4.5 million years. Monthly global CO_2 concentrations exceeded 400 ppm in March 2015, probably for the first time in several million years. On 12 November 2015, NASA scientists reported that human-made carbon dioxide continues to increase above levels not seen in hundreds of thousands of years: currently, about half of the carbon dioxide released from the burning of fossil fuels is not absorbed by vegetation and the oceans and remains in the atmosphere.

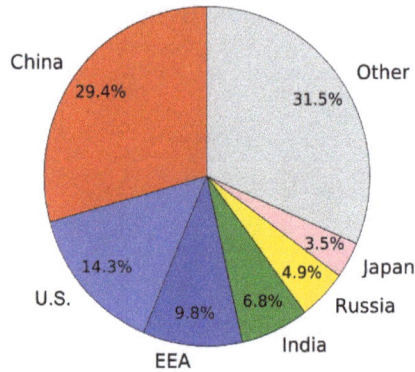

Global carbon dioxide emissions by country.

Over the last three decades of the twentieth century, gross domestic product per capita and population growth were the main drivers of increases in greenhouse gas emissions. CO_2 emissions are continuing to rise due to the burning of fossil fuels and land-use change. Emissions can be attributed to different regions. Attributions of emissions due to land-use change are subject to considerable uncertainty.

Emissions scenarios, estimates of changes in future emission levels of greenhouse gases, have been projected that depend upon uncertain economic, sociological, technological, and natural developments. In most scenarios, emissions continue to rise over the century, while in a few, emissions are reduced. Fossil fuel reserves are abundant, and will not limit carbon emissions in the 21st century. Emission scenarios, combined with modelling of the carbon cycle, have been used to produce estimates of how atmospheric concentrations of greenhouse gases might change in the future. Using the six IPCC SRES "marker" scenarios, models suggest that by the year 2100, the atmospheric concentration of CO_2 could range between 541 and 970 ppm. This is 90–250% above the concentration in the year 1750.

The popular media and the public often confuse global warming with ozone depletion, i.e., the destruction of stratospheric ozone (e.g., the ozone layer) by chlorofluorocarbons. Although there are a few areas of linkage, the relationship between the two is not strong. Reduced stratospheric ozone has had a slight cooling influence on surface temperatures, while increased tropospheric ozone has had a somewhat larger warming effect.

Aerosols and Soot

Global dimming, a gradual reduction in the amount of global direct irradiance at the Earth's surface, was observed from 1961 until at least 1990. Solid and liquid particles known as *aerosols*, produced by volcanoes and human-made pollutants, are thought to be the main cause of this dimming. They exert a cooling effect by increasing the reflection of incoming sunlight. The effects of the products of fossil fuel combustion – CO_2 and aerosols – have partially offset one another in recent decades, so that net warming has been due to the increase in non-CO_2 greenhouse gases such as methane. Radiative forcing due to aerosols is temporally limited due to the processes that remove aerosols from the atmosphere. Removal by clouds and precipitation gives tropospheric aerosols an atmospheric lifetime of only about a week, while stratospheric aerosols can remain for a few years. Carbon dioxide has a lifetime of a century or more, and as such, changes in aerosols will only delay climate changes due to carbon dioxide. Black carbon is second only to carbon dioxide for its contribution to global warming (contribution being estimated at 17 to 20%, whereas carbon dioxide contributes 40 to 45% to global warming).

Ship tracks can be seen as lines in these clouds over the Atlantic Ocean on the east coast of the United States. Atmospheric particles from these and other sources could have a large effect on climate through the aerosol indirect effect.

In addition to their direct effect by scattering and absorbing solar radiation, aerosols have indirect effects on the Earth's radiation budget. Sulfate aerosols act as cloud condensation nuclei and thus lead to clouds that have more and smaller cloud droplets. These clouds reflect solar radiation more efficiently than clouds with fewer and larger droplets, a phenomenon known as the Twomey effect. This effect also causes droplets to be of more uniform size, which reduces growth of raindrops and makes the cloud more reflective to incoming sunlight, known as the Albrecht effect. Indirect effects are most noticeable in marine stratiform clouds, and have very little radiative effect on convective clouds. Indirect effects of aerosols represent the largest uncertainty in radiative forcing.

Soot may either cool or warm Earth's climate system, depending on whether it is airborne or deposited. Atmospheric soot directly absorbs solar radiation, which heats the atmosphere and cools the surface. In isolated areas with high soot production, such as rural India, as much as 50% of surface warming due to greenhouse gases may be masked by atmospheric brown clouds. When deposited, especially on glaciers or on ice in arctic regions, the lower surface albedo can also directly heat the surface. The influences of atmospheric particles, including black carbon, are most pronounced in the tropics and sub-tropics, particularly in Asia, while the effects of greenhouse gases are dominant in the extratropics and southern hemisphere.

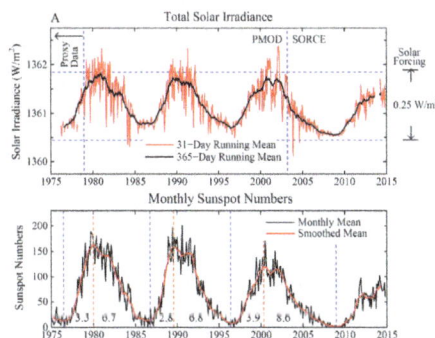

Changes in total solar irradiance (TSI) and monthly sunspot numbers since the mid-1970s.

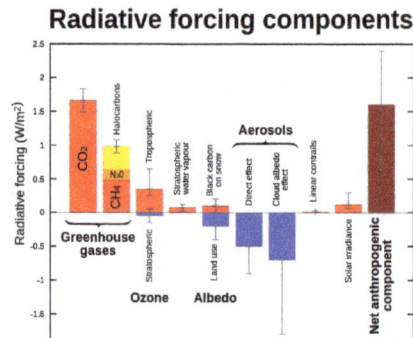

Contribution of natural factors and human activities to radiative forcing of climate change.
Radiative forcing values are for the year 2005, relative to the pre-industrial era (1750). The contribution of
solar irradiance to radiative forcing is 5% the value of the combined radiative forcing due to increases in the
atmospheric concentrations of carbon dioxide, methane and nitrous oxide.

Solar Activity

Since 1978, solar irradiance has been measured by satellites. These measurements indicate that
the Sun's radiative output has not increased during that time, so the warming during the past 40
years cannot be attributed to an increase in solar energy reaching the Earth.

Climate models have been used to examine the role of the Sun in recent climate change. Models
are unable to reproduce the rapid warming observed in recent decades when they only take into
account variations in solar output and volcanic activity. Models are, however, able to simulate the
observed 20th century changes in temperature when they include all of the most important external forcings, including human influences and natural forcings.

Another line of evidence is differing temperature changes at different levels in the Earth's atmosphere. Basic physical principles require that the greenhouse effect produces warming of the lower
atmosphere (the troposphere) but cooling of the upper atmosphere (the stratosphere). Depletion
of the ozone layer by chemical refrigerants has also resulted in a strong cooling effect in the stratosphere. If solar variations were responsible for observed warming, warming of both the troposphere and stratosphere would be expected.

Variations in Earth's orbit

The tilt of the Earth's axis and the shape of its orbit around the Sun vary slowly over tens of thousands of years. This changes climate by changing the seasonal and latitudinal distribution of incoming solar energy at Earth's surface. During the last few thousand years, this phenomenon contributed to a slow cooling trend at high latitudes of the Northern Hemisphere during summer, a
trend that was reversed by greenhouse-gas-induced warming during the 20th century. Orbital
cycles favorable for glaciation are not expected within the next 50,000 years.

Feedback

The climate system includes a range of *feedbacks*, which alter the response of the system to changes in external forcings. Positive feedbacks increase the response of the climate system to an initial
forcing, while negative feedbacks reduce it.

Sea ice, shown here in Nunavut, in northern Canada, reflects more sunshine, while open ocean absorbs more, accelerating melting.

There are a range of feedbacks in the climate system, including water vapour, changes in ice-albedo (snow and ice cover affect how much the Earth's surface absorbs or reflects incoming sunlight), clouds, and changes in the Earth's carbon cycle (e.g., the release of carbon from soil). The main negative feedback is the energy the Earth's surface radiates into space as infrared radiation. According to the Stefan-Boltzmann law, if the absolute temperature (as measured in kelvins) doubles, radiated energy increases by a factor of 16 (2 to the 4th power).

Feedbacks are an important factor in determining the sensitivity of the climate system to increased atmospheric greenhouse gas concentrations. Other factors being equal, a higher *climate sensitivity* means that more warming will occur for a given increase in greenhouse gas forcing. Uncertainty over the effect of feedbacks is a major reason why different climate models project different magnitudes of warming for a given forcing scenario. More research is needed to understand the role of clouds and carbon cycle feedbacks in climate projections.

The IPCC projections previously mentioned span the "likely" range (greater than 66% probability, based on expert judgement) for the selected emissions scenarios. However, the IPCC's projections do not reflect the full range of uncertainty. The lower end of the "likely" range appears to be better constrained than the upper end.

Climate Models

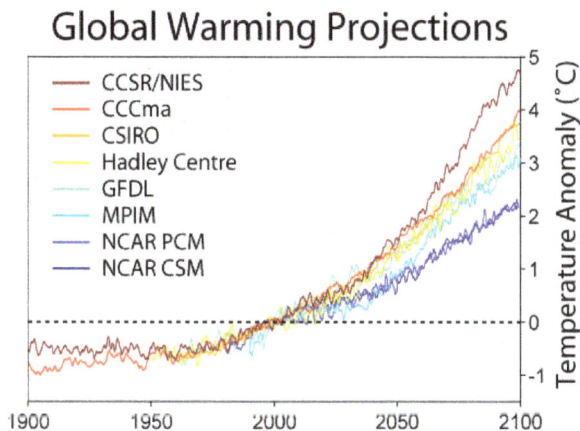

Calculations of global warming prepared in or before 2001 from a range of climate models under the SRES A2 emissions scenario, which assumes no action is taken to reduce emissions and regionally divided economic development.

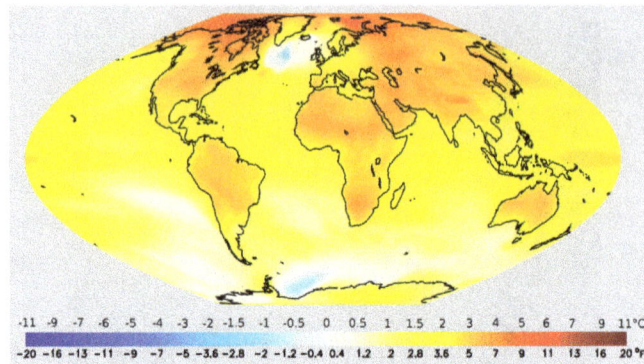

Projected change in annual mean surface air temperature from the late 20th century to the middle 21st century, based on a medium emissions scenario (SRES A1B). This scenario assumes that no future policies are adopted to limit greenhouse gas emissions.

A climate model is a representation of the physical, chemical and biological processes that affect the climate system. Such models are based on scientific disciplines such as fluid dynamics and thermodynamics as well as physical processes such as radiative transfer. The models may be used to predict a range of variables such as local air movement, temperature, clouds, and other atmospheric properties; ocean temperature, salt content, and circulation; ice cover on land and sea; the transfer of heat and moisture from soil and vegetation to the atmosphere; and chemical and biological processes, among others.

Although researchers attempt to include as many processes as possible, simplifications of the actual climate system are inevitable because of the constraints of available computer power and limitations in knowledge of the climate system. Results from models can also vary due to different greenhouse gas inputs and the model's climate sensitivity. For example, the uncertainty in IPCC's 2007 projections is caused by (1) the use of multiple models with differing sensitivity to greenhouse gas concentrations, (2) the use of differing estimates of humanity's future greenhouse gas emissions, (3) any additional emissions from climate feedbacks that were not included in the models IPCC used to prepare its report, i.e., greenhouse gas releases from permafrost.

The models do not assume the climate will warm due to increasing levels of greenhouse gases. Instead the models predict how greenhouse gases will interact with radiative transfer and other physical processes. Warming or cooling is thus a result, not an assumption, of the models.

Clouds and their effects are especially difficult to predict. Improving the models' representation of clouds is therefore an important topic in current research. Another prominent research topic is expanding and improving representations of the carbon cycle.

Models are also used to help investigate the causes of recent climate change by comparing the observed changes to those that the models project from various natural and human causes. Although these models do not unambiguously attribute the warming that occurred from approximately 1910 to 1945 to either natural variation or human effects, they do indicate that the warming since 1970 is dominated by anthropogenic greenhouse gas emissions.

The physical realism of models is tested by examining their ability to simulate contemporary or past climates. Climate models produce a good match to observations of global temperature changes over the last century, but do not simulate all aspects of climate. Not all effects of global warming are accurately predicted by the climate models used by the IPCC. Observed Arctic shrinkage has been faster than that predicted. Precipitation increased proportionally to atmospheric humidity, and hence significantly faster than global climate models predict. Since 1990, sea level has also risen considerably faster than models predicted it would.

Observed and Expected Environmental Effects

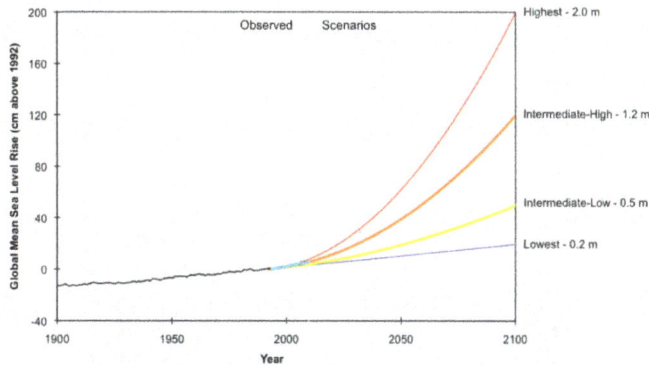

Projections of global mean sea level rise by Parris and others. Probabilities have not been assigned to these projections. Therefore, none of these projections should be interpreted as a "best estimate" of future sea level rise.

Anthropogenic forcing has likely contributed to some of the observed changes, including sea level rise, changes in climate extremes (such as the number of warm and cold days), declines in Arctic sea ice extent, glacier retreat, and greening of the Sahara.

During the 21st century, glaciers and snow cover are projected to continue their widespread retreat. Projections of declines in Arctic sea ice vary. Recent projections suggest that Arctic summers could be ice-free (defined as ice extent less than 1 million square km) as early as 2025–2030.

"Detection" is the process of demonstrating that climate has changed in some defined statistical sense, without providing a reason for that change. Detection does not imply attribution of the detected change to a particular cause. "Attribution" of causes of climate change is the process of establishing the most likely causes for the detected change with some defined level of confidence. Detection and attribution may also be applied to observed changes in physical, ecological and social systems.

Extreme Weather

Changes in regional climate are expected to include greater warming over land, with most warming at high northern latitudes, and least warming over the Southern Ocean and parts of the North Atlantic Ocean.

Future changes in precipitation are expected to follow existing trends, with reduced precipitation over subtropical land areas, and increased precipitation at subpolar latitudes and some equatorial regions. Projections suggest a probable increase in the frequency and severity of some extreme weather events, such as heat waves.

A 2015 study published in *Nature Climate Change*, states:

About 18% of the moderate daily precipitation extremes over land are attributable to the observed temperature increase since pre-industrial times, which in turn primarily results from human influence. For 2 °C of warming the fraction of precipitation extremes attributable to human influence rises to about 40%. Likewise, today about 75% of the moderate daily hot extremes over land are attributable to warming. It is the most rare and extreme events for which the largest fraction is anthropogenic, and that contribution increases nonlinearly with further warming.

Data analysis of extreme events from 1960 until 2010 suggests that droughts and heat waves appear simultaneously with increased frequency. Extremely wet or dry events within the monsoon period have increased since 1980.

Sea level Rise

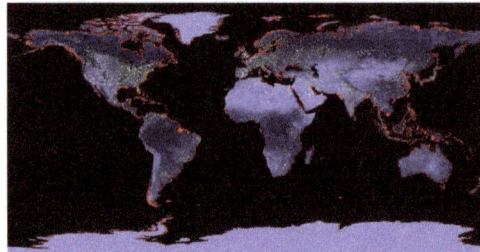

Map of the Earth with a six-metre sea level rise represented in red.

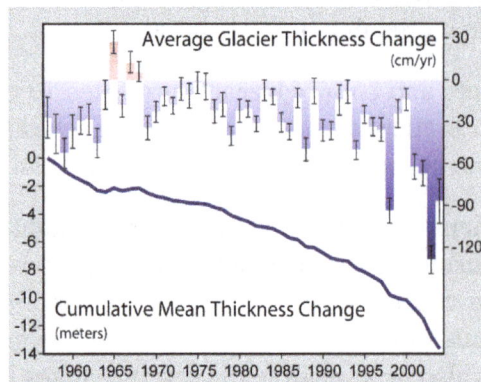

Sparse records indicate that glaciers have been retreating since the early 1800s. In the 1950s measurements began that allow the monitoring of glacial mass balance, reported to the World Glacier Monitoring Service (WGMS) and the National Snow and Ice Data Center (NSIDC).

The sea level rise since 1993 has been estimated to have been on average 2.6 mm and 2.9 mm per year ± 0.4 mm. Additionally, sea level rise has accelerated from 1995 to 2015. Over the 21st century, the IPCC projects for a high emissions scenario, that global mean sea level could rise by 52–98 cm. The IPCC's projections are conservative, and may underestimate future sea level rise. Other estimates suggest that for the same period, global mean sea level could rise by 0.2 to 2.0 m (0.7–6.6 ft), relative to mean sea level in 1992.

Widespread coastal flooding would be expected if several degrees of warming is sustained for millennia. For example, sustained global warming of more than 2 °C (relative to pre-industrial levels) could lead to eventual sea level rise of around 1 to 4 m due to thermal expansion of sea water and the melting of glaciers and small ice caps. Melting of the Greenland ice sheet could contribute an additional 4 to 7.5 m over many thousands of years. It has been estimated that we are already committed to a sea-level rise of approximately 2.3 metres for each degree of temperature rise within the next 2,000 years.

Warming beyond the 2 °C target would potentially lead to rates of sea-level rise dominated by ice loss from Antarctica. Continued CO_2 emissions from fossil sources could cause additional tens of metres of sea level rise, over the next millennia and eventually ultimately eliminate the entire Antarctic ice sheet, causing about 58 metres of sea level rise.

Ecological Systems

In terrestrial ecosystems, the earlier timing of spring events, as well as poleward and upward shifts in plant and animal ranges, have been linked with high confidence to recent warming. Future climate change is expected to affect particular ecosystems, including tundra, mangroves, coral reefs, and caves. It is expected that most ecosystems will be affected by higher atmospheric CO_2 levels, combined with higher global temperatures. Overall, it is expected that climate change will result in the extinction of many species and reduced diversity of ecosystems.

Increases in atmospheric CO_2 concentrations have led to an increase in ocean acidity. Dissolved CO_2 increases ocean acidity, measured by lower pH values. Between 1750 and 2000, surface-ocean pH has decreased by ≈0.1, from ≈8.2 to ≈8.1. Surface-ocean pH has probably not been below ≈8.1 during the past 2 million years. Projections suggest that surface-ocean pH could decrease by an additional 0.3–0.4 units by 2100. Future ocean acidification could threaten coral reefs, fisheries, protected species, and other natural resources of value to society.

Ocean deoxygenation is projected to increase hypoxia by 10%, and triple suboxic waters (oxygen concentrations 98% less than the mean surface concentrations), for each 1 °C of upper ocean warming.

Long-term Effects

On the timescale of centuries to millennia, the magnitude of global warming will be determined primarily by anthropogenic CO_2 emissions. This is due to carbon dioxide's very long lifetime in the atmosphere.

Stabilizing the global average temperature would require large reductions in CO_2 emissions, as well as reductions in emissions of other greenhouse gases such as methane and nitrous oxide. Emissions of CO_2 would need to be reduced by more than 80% relative to their peak level. Even if this were achieved, global average temperatures would remain close to their highest level for many centuries. As of 2016, emissions of CO_2 from burning fossil fuels had stopped increasing, but the Guardian reports they need to be "reduced to have a real impact on climate change". Meanwhile, this greenhouse gas continues to accumulate in the atmosphere. Also, CO_2 is not the only factor driving climate change. Concentrations of atmospheric methane, another greenhouse gas, rose dramatically between 2006–2016 for unknown reasons. This undermines efforts to combat global warming and there is a risk of an uncontrollable runaway greenhouse effect.

Long-term effects also include a response from the Earth's crust, due to ice melting and deglaciation, in a process called post-glacial rebound, when land masses are no longer depressed by the weight of ice. This could lead to landslides and increased seismic and volcanic activities. Tsunamis could be generated by submarine landslides caused by warmer ocean water thawing ocean-floor permafrost or releasing gas hydrates. Some world regions, such as the French Alps, already show signs of an increase in landslide frequency.

Large-scale and Abrupt Impacts

Climate change could result in global, large-scale changes in natural and social systems. Examples include the possibility for the Atlantic Meridional Overturning Circulation to slow- or shutdown, which in the instance of a shutdown would change weather in Europe and North America considerably, ocean acidification caused by increased atmospheric concentrations of carbon dioxide, and the long-term melting of ice sheets, which contributes to sea level rise.

Some large-scale changes could occur abruptly, i.e., over a short time period, and might also be irreversible. Examples of abrupt climate change are the rapid release of methane and carbon dioxide from permafrost, which would lead to amplified global warming, or the shutdown of thermohaline circulation. Scientific understanding of abrupt climate change is generally poor. The probability of abrupt change for some climate related feedbacks may be low. Factors that may increase the probability of abrupt climate change include higher magnitudes of global warming, warming that occurs more rapidly, and warming that is sustained over longer time periods.

Observed and Expected Effects on Social Systems

The effects of climate change on human systems, mostly due to warming or shifts in precipitation patterns, or both, have been detected worldwide. Production of wheat and maize globally has been impacted by climate change. While crop production has increased in some mid-latitude regions such as the UK and Northeast China, economic losses due to extreme weather events have increased globally. There has been a shift from cold- to heat-related mortality in some regions as a result of warming. Livelihoods of indigenous peoples of the Arctic have been altered by climate change, and there is emerging evidence of climate change impacts on livelihoods of indigenous peoples in other regions. Regional impacts of climate change are now observable at more locations than before, on all continents and across ocean regions.

The future social impacts of climate change will be uneven. Many risks are expected to increase with higher magnitudes of global warming. All regions are at risk of experiencing negative impacts. Low-latitude, less developed areas face the greatest risk. A study from 2015 concluded that economic growth (gross domestic product) of poorer countries is much more impaired with projected future climate warming, than previously thought.

A meta-analysis of 56 studies concluded in 2014 that each degree of temperature rise will increase violence by up to 20%, which includes fist fights, violent crimes, civil unrest or wars.

Examples of impacts include:

Food: Crop production will probably be negatively affected in low latitude countries, while effects at northern latitudes may be positive or negative. Global warming of around 4.6 °C relative to pre-industrial levels could pose a large risk to global and regional food security.

Health: Generally impacts will be more negative than positive. Impacts include: the effects of extreme weather, leading to injury and loss of life; and indirect effects, such as undernutrition brought on by crop failures.

Habitat Inundation

In small islands and mega deltas, inundation as a result of sea level rise is expected to threaten vital infrastructure and human settlements. This could lead to issues of homelessness in countries with low-lying areas such as Bangladesh, as well as statelessness for populations in countries such as the Maldives and Tuvalu.

Economy

Estimates based on the IPCC A1B emission scenario from additional CO_2 and CH_4 greenhouse gases released from permafrost, estimate associated impact damages by US$43 trillion.

Infrastructure

Continued permafrost degradation will likely result in unstable infrastructure in Arctic regions, or Alaska before 2100. Thus, impacting roads, pipelines and buildings, as well as water distribution, and cause slope failures.

Possible Responses to Global Warming

Mitigation

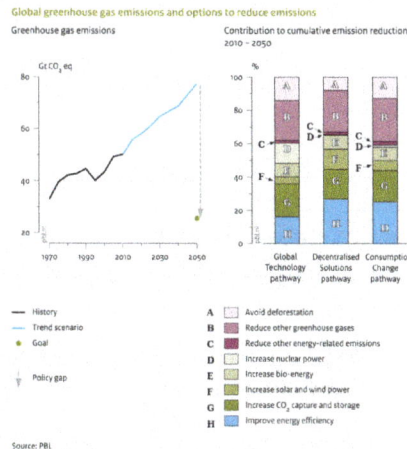

The graph on the right shows three "pathways" to meet the UNFCCC's 2 °C target, labelled "global technology", "decentralized solutions", and "consumption change". Each pathway shows how various measures (e.g., improved energy efficiency, increased use of renewable energy) could contribute to emissions reductions.

Mitigation of climate change are actions to reduce greenhouse gas emissions, or enhance the capacity of carbon sinks to absorb GHGs from the atmosphere. There is a large potential for future reductions in emissions by a combination of activities, including: energy conservation and increased energy efficiency; the use of low-carbon energy technologies, such as renewable energy, nuclear energy, and carbon capture and storage; and enhancing carbon sinks through, for example, reforestation and preventing deforestation. A 2015 report by Citibank concluded that transitioning to a low carbon economy would yield positive return on investments.

Near- and long-term trends in the global energy system are inconsistent with limiting global warming at below 1.5 or 2 °C, relative to pre-industrial levels. Pledges made as part of the Cancún agreements are broadly consistent with having a likely chance (66 to 100% probability) of limiting global warming (in the 21st century) at below 3 °C, relative to pre-industrial levels.

In limiting warming at below 2 °C, more stringent emission reductions in the near-term would allow for less rapid reductions after 2030. Many integrated models are unable to meet the 2 °C target if pessimistic assumptions are made about the availability of mitigation technologies.

Adaptation

Other policy responses include adaptation to climate change. Adaptation to climate change may be planned, either in reaction to or anticipation of climate change, or spontaneous, i.e., without government intervention. Planned adaptation is already occurring on a limited basis. The barriers, limits, and costs of future adaptation are not fully understood.

A concept related to adaptation is *adaptive capacity*, which is the ability of a system (human, natural or managed) to adjust to climate change (including climate variability and extremes) to moderate potential damages, to take advantage of opportunities, or to cope with consequences. Unmitigated climate change (i.e., future climate change without efforts to limit greenhouse gas emissions) would, in the long term, be likely to exceed the capacity of natural, managed and human systems to adapt.

Environmental organizations and public figures have emphasized changes in the climate and the risks they entail, while promoting adaptation to changes in infrastructural needs and emissions reductions.

Climate Engineering

Climate engineering (sometimes called *geoengineering* or *climate intervention*) is the deliberate modification of the climate. It has been investigated as a possible response to global warming, e.g. by NASA and the Royal Society. Techniques under research fall generally into the categories solar radiation management and carbon dioxide removal, although various other schemes have been suggested. A study from 2014 investigated the most common climate engineering methods and concluded they are either ineffective or have potentially severe side effects and cannot be stopped without causing rapid climate change.

Discourse about Global Warming

Political Discussion

Most countries in the world are parties to the United Nations Framework Convention on Climate Change (UNFCCC). The ultimate objective of the Convention is to prevent dangerous human interference of the climate system. As stated in the Convention, this requires that GHG concentrations are stabilized in the atmosphere at a level where ecosystems can adapt naturally to climate change, food production is not threatened, and economic development can proceed in a sustainable fashion. The Framework Convention was agreed in 1992, but since then, global emissions have risen.

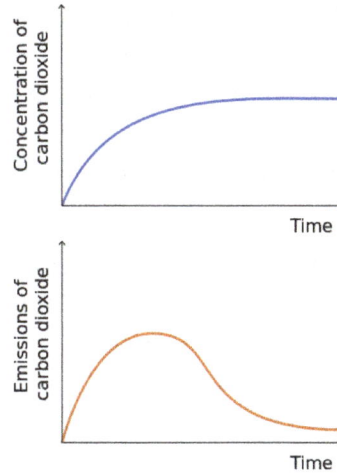

Article 2 of the UN Framework Convention refers explicitly to "stabilization of greenhouse gas concentrations." To stabilize the atmospheric concentration of CO_2, emissions worldwide would need to be dramatically reduced from their present level.

During negotiations, the G77 (a lobbying group in the United Nations representing 133 developing nations) pushed for a mandate requiring developed countries to "[take] the lead" in reducing their emissions. This was justified on the basis that: the developed world's emissions had contributed most to the cumulation of GHGs in the atmosphere; per-capita emissions (i.e., emissions per head of population) were still relatively low in developing countries; and the emissions of developing countries would grow to meet their development needs.

This mandate was sustained in the Kyoto Protocol to the Framework Convention, which entered into legal effect in 2005. In ratifying the Kyoto Protocol, most developed countries accepted legally binding commitments to limit their emissions. These first-round commitments expired in 2012. United States President George W. Bush rejected the treaty on the basis that "it exempts 80% of the world, including major population centres such as China and India, from compliance, and would cause serious harm to the US economy."

At the 15th UNFCCC Conference of the Parties, held in 2009 at Copenhagen, several UNFCCC Parties produced the Copenhagen Accord. Parties associated with the Accord (140 countries, as of November 2010) aim to limit the future increase in global mean temperature to below 2 °C. The 16th Conference of the Parties (COP16) was held at Cancún in 2010. It produced an agreement, not a binding treaty, that the Parties should take urgent action to reduce greenhouse gas emissions to meet a goal of limiting global warming to 2 °C above pre-industrial temperatures. It also recognized the need to consider strengthening the goal to a global average rise of 1.5 °C.

Scientific Discussion

There is continuing discussion through published peer-reviewed scientific papers, which are assessed by scientists working in the relevant fields taking part in the Intergovernmental Panel on Climate Change. The scientific consensus as of 2013 stated in the IPCC Fifth Assessment Report is that it "is extremely likely that human influence has been the dominant cause of the observed warming since the mid-20th century". A 2008 report by the U.S. National Academy of Sciences stated that most scientists by then agreed that observed warming in recent decades was primarily caused by human activities increasing the amount of greenhouse gases in the atmosphere. In 2005 the Royal Society stated that while the overwhelming majority of scientists were in agreement on the main points, some individuals and organizations opposed to the consensus on urgent action needed to reduce greenhouse gas emissions had tried to undermine the science and work of the IPCC. National science academies have called on world leaders for policies to cut global emissions.

In the scientific literature, there is a strong consensus that global surface temperatures have increased in recent decades and that the trend is caused mainly by human-induced emissions of greenhouse gases. No scientific body of national or international standing disagrees with this view.

Discussion by the Public and in Popular Media

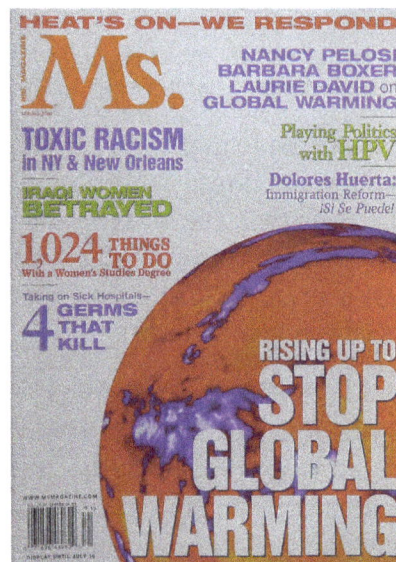

Global warming was the cover story in this 2007 issue of *Ms.* magazine

The global warming controversy refers to a variety of disputes, substantially more pronounced in the popular media than in the scientific literature, regarding the nature, causes, and consequences of global warming. The disputed issues include the causes of increased global average air temperature, especially since the mid-20th century, whether this warming trend is unprecedented or within normal climatic variations, whether humankind has contributed significantly to it, and whether the increase is completely or partially an artefact of poor measurements. Additional disputes concern estimates of climate sensitivity, predictions of additional warming, and what the consequences of global warming will be.

By 1990, American conservative think tanks had begun challenging the legitimacy of global warming as a social problem. They challenged the scientific evidence, argued that global warming would have benefits, and asserted that proposed solutions would do more harm than good. Some people dispute aspects of climate change science. Organizations such as the libertarian Competitive Enterprise Institute, conservative commentators, and some companies such as ExxonMobil have challenged IPCC climate change scenarios, funded scientists who disagree with the scientific consensus, and provided their own projections of the economic cost of stricter controls. On the other hand, some fossil fuel companies have scaled back their efforts in recent years, or even called for policies to reduce global warming. Global oil companies have begun to acknowledge climate change exists and is caused by human activities and the burning of fossil fuels.

Surveys of Public Opinion

The world public, or at least people in economically advanced regions, became broadly aware of the global warming problem in the late 1980s. Polling groups began to track opinions on the subject, at first mainly in the United States. The longest consistent polling, by Gallup in the US, found relatively small deviations of 10% or so from 1998 to 2015 in opinion on the seriousness of global warming, but with increasing polarization between those concerned and those unconcerned.

The first major worldwide poll, conducted by Gallup in 2008–2009 in 127 countries, found that some 62% of people worldwide said they knew about global warming. In the advanced countries of North America, Europe and Japan, 90% or more knew about it (97% in the U.S., 99% in Japan); in less developed countries, especially in Africa, fewer than a quarter knew about it, although many had noticed local weather changes. Among those who knew about global warming, there was a wide variation between nations in belief that the warming was a result of human activities.

By 2010, with 111 countries surveyed, Gallup determined that there had been a substantial decrease since 2007–2008 in the number of Americans and Europeans who viewed global warming as a serious threat. In the US, just a little over half the population (53%) viewed it as a serious concern for either themselves or their families; this was 10 points below the 2008 poll (63%). Latin America had the biggest rise in concern: 73% said global warming was a serious threat to their families. This global poll also found that people were more likely to attribute global warming to human activities than to natural causes, except in the US where nearly half (47%) of the population attributed global warming to natural causes.

A March–May 2013 survey by Pew Research Center for the People & the Press polled 39 countries about global threats. According to 54% of those questioned, global warming featured top of the perceived global threats. In a January 2013 survey, Pew found that 69% of Americans say there is solid evidence that the Earth's average temperature has gotten warmer over the past few decades, up six points since November 2011 and 12 points since 2009.

A 2010 survey of 14 industrialized countries found that skepticism about the danger of global warming was highest in Australia, Norway, New Zealand and the United States, in that order, correlating positively with per capita emissions of carbon dioxide.

Etymology

In the 1950s, research suggested increasing temperatures, and a 1952 newspaper reported "climate change". This phrase next appeared in a November 1957 report in *The Hammond Times* which described Roger Revelle's research into the effects of increasing human-caused CO_2 emissions on the greenhouse effect, "a large scale global warming, with radical climate changes may result". Both phrases were only used occasionally until 1975, when Wallace Smith Broecker published a scientific paper on the topic, "Climatic Change: Are We on the Brink of a Pronounced Global Warming?" The phrase began to come into common use, and in 1976 Mikhail Budyko's statement that "a global warming up has started" was widely reported. Other studies, such as a 1971 MIT report, referred to the human impact as "inadvertent climate modification", but an influential 1979 National Academy of Sciences study headed by Jule Charney followed Broecker in using *global warming* for rising surface temperatures, while describing the wider effects of increased CO_2 as *climate change*.

In 1986 and November 1987, NASA climate scientist James Hansen gave testimony to Congress on global warming. There were increasing heatwaves and drought problems in the summer of 1988, and when Hansen testified in the Senate on 23 June he sparked worldwide interest. He said: "global warming has reached a level such that we can ascribe with a high degree of confidence a cause and effect relationship between the greenhouse effect and the observed warming." Public attention increased over the summer, and *global warming* became the dominant popular term, commonly used both by the press and in public discourse.

In a 2008 NASA article on usage, Erik M. Conway defined *global warming* as "the increase in Earth's average surface temperature due to rising levels of greenhouse gases", while *climate change* was "a long-term change in the Earth's climate, or of a region on Earth." As effects such as changing patterns of rainfall and rising sea levels would probably have more impact than temperatures alone, he considered *global climate change* a more scientifically accurate term, and like the Intergovernmental Panel on Climate Change, the NASA website would emphasize this wider context.

Arctic Sea Ice Decline

September 2, 2012 (The record lowest minimum ever observed in the satellite record occurred on September 16, 2012, when sea ice declined to 3.41 million square kilometers (1.32 million square miles). This image shows the area two weeks earlier). Satellite views of Arctic sea ice.

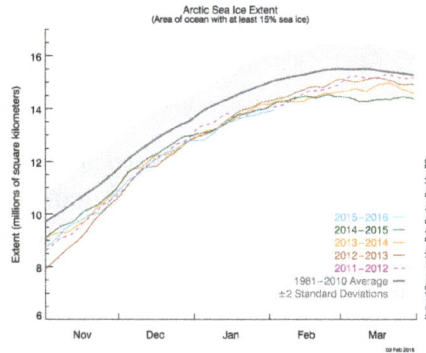

Arctic sea ice extent as of February 3, 2016. January Arctic sea ice extent was the lowest in the satellite record.

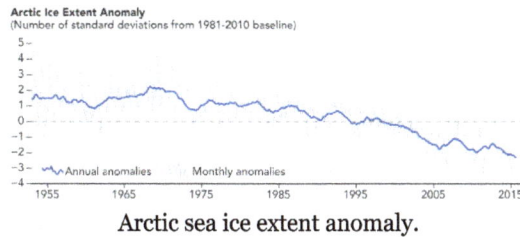

Arctic sea ice extent anomaly.

Arctic sea ice decline is the sea ice loss observed in recent decades in the Arctic Ocean. The Intergovernmental Panel on Climate Change (IPCC) Fourth Assessment Report states that greenhouse gas forcing is largely, but not wholly, responsible for the decline in Arctic sea ice extent. A study from 2011 suggested that internal variability enhanced the greenhouse gas forced sea ice decline over the last decades. A study from 2007 found the decline to be "faster than forecasted" by model simulations. The IPCC Fifth Assessment Report concluded with high confidence that sea ice continues to decrease in extent, and that there is robust evidence for the downward trend in Arctic summer sea ice extent since 1979. It has been established that the region is at its warmest for at least 40,000 years and the Arctic-wide melt season has lengthened at a rate of 5 days per decade (from 1979 to 2013), dominated by a later autumn freezeup. Sea ice changes have been identified as a mechanism for polar amplification.

Definitions

The Arctic Ocean is the mass of water positioned approximately above latitude 65° N. Arctic Sea Ice refers to the area of the Arctic Ocean covered in ice. This value fluctuates during the yearly period and overall is declining rapidly. The Arctic sea ice minimum is the day in a given year when Arctic sea ice reaches its smallest extent. It occurs at the end of the summer melting season, normally during September. Arctic Sea ice maximum is the day of a year when Arctic sea ice reaches its largest extent near the end of the Arctic cold season, normally during March. Typical data visualizations for Arctic sea ice include average monthly measurements or graphs for the annual minimum or maximum extent, as shown in the images above.

Observation

Observation with satellites show that Arctic sea ice area, extent, and volume have been in decline for a few decades. Sometime during the 21st century, sea ice may effectively cease to exist during the summer. Sea ice extent is defined as the area with at least 15% ice cover. The amount of multi-year sea ice in the Arctic has declined considerably in recent decades. In 1988, ice that was at least 4 years old accounted for 26% of the Arctic's sea ice. By 2013, ice that age was only 7% of all Arctic sea ice.

Scientists recently measured sixteen-foot (five-meter) wave heights during a storm in the Beaufort Sea in mid-August until late October 2012. This is a new phenomenon for the region, since a permanent sea ice cover normally prevents wave formation. Wave action breaks up sea ice, and thus could become a feedback mechanism, driving sea ice decline.

For January 2016, the satellite based data showed the lowest overall Arctic sea ice extent of any January since records begun in 1979. Bob Henson from Wunderground noted:

Hand in hand with the skimpy ice cover, temperatures across the Arctic have been extraordinarily warm for midwinter. Just before New Year's, a slug of mild air pushed temperatures above freezing to within 200 miles of the North Pole. That warm pulse quickly dissipated, but it was followed by a series of intense North Atlantic cyclones that sent very mild air poleward, in tandem with a strongly negative Arctic Oscillation during the first three weeks of the month.

The January 2016's remarkable phase transition of Arctic Oscillation was driven by a rapid tropospheric warming in the Arctic, a pattern that appears to have increased surpassing the so-called stratospheric sudden warming. The previous record of the lowest area of the Arctic Ocean covered by ice in 2012 saw a low of 1.58 million square miles (4.1 million square kilometers). This replaced the previous record set on September 28, 2007 at 1.61 million square miles(4.17 million square kilometers).

Ice-free Summer

As ice melts, the liquid water collects in depressions on the surface and deepens them, forming these melt ponds in the Arctic. These fresh water ponds are separated from the salty sea below and around it, until breaks in the ice merge the two.

An "ice-free" Arctic Ocean is often defined as "having less than 1 million square kilometers of sea ice", because it is very difficult to melt the thick ice around the Canadian Arctic Archipelago. The IPCC AR5 defines "nearly ice-free conditions" as sea ice extent less than 10^6 km^2 for at least five consecutive years.

Many scientists have attempted to estimate when the Arctic will be "ice-free". Professor Peter Wadhams of the University of Cambridge is among these scientists. Wadhams and several others have noted that climate model predictions have been overly conservative regarding sea ice decline. A 2013 paper suggested that models commonly underestimate the solar radiation absorption characteristics of wildfire soot. A 2006 paper predicted "near ice-free September conditions by 2040". Overland & Wang (2009) predicted that there would be an ice-free Arctic in the summer by 2037. The same year Boé et al. found that the Arctic will probably be ice-free in September before the end of the 21st century. A follow-up study concluded with the possibility of major sea ice loss within a decade or two. The IPCC AR5 (for at least one scenario) estimates an ice-free summer might occur around 2050. The Third U.S. National Climate Assessment (NCA), released May 6, 2014, reports that the Arctic Ocean is expected to be ice free in summer before mid-century. Models that best match historical trends project a nearly ice-free Arctic in the summer by the 2030s. However, these models do tend to underestimate the rate of sea ice loss since 2007. A 2010 study suggested that the Arctic Ocean will be ice-free sooner than global climate models predict. They chart the summer of 2016 as ice-free, but show a possible date range out to 2020. This assessment was reported in the press as "US Navy predicts summer ice free Arctic by 2016." In a study from 2016, the prediction uncertainty of an ice-free Arctic was quantified to be at around two decades, based on model simulations.

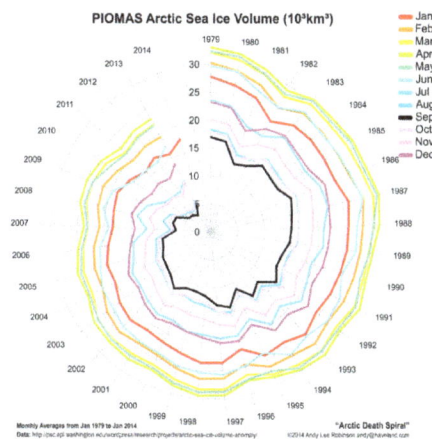

Monthly averages from January 1979 - January 2014. Data source via the Polar Science Center (University of Washington).

Tipping Point

There is an ongoing debate if the Arctic Ocean has already passed a "tipping point", defined as a threshold for abrupt and irreversible change. A 2013 study identified an abrupt transition to increased seasonal ice cover variability in 2007, which has persisted in following years. The researchers made a distinction between a bifurcation and a non-bifurcation `tipping point'. The *IPCC AR5* report stated with medium confidence that precise levels of climate change sufficient to trigger a tipping point remain uncertain, and that the risk associated with crossing multiple tipping points increases with rising temperature.

Implications

Physical implications which arise from lesser ocean surface covered sea-ice include the ice-albedo feedback or warmer sea surface temperatures which increase ocean heat content. As Peter Wadhams, a polar researcher writes "once summer ice yields to open water, the albedo... drops from

0.6 to 0.1, which will further accelerate warming of the Arctic and of the whole planet" This in turn changes evaporation patterns and the polar vortex. Economical implications of ice free summers and the decline in Arctic ice volumes include a greater number of journeys across the Arctic Ocean Shipping lanes during the year. This number has grown from 0, in 1979 to 400-500 along the Bering strait and >40 along the Northern Sea Route, in 2013.

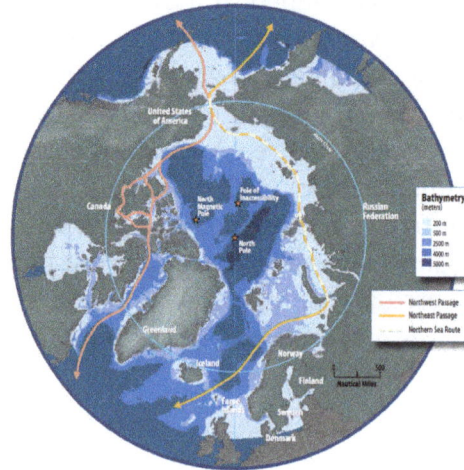

Map illustrating various Arctic shipping routes

Atmospheric Chemistry

Melting of sea ice releases molecular chlorine, which reacts with sunlight to produce chlorine atoms. Because chlorine atoms are highly reactive, they can expedite the degradation of methane and tropospheric ozone and the oxidation of mercury to more toxic forms. Cracks in sea ice are causing ozone and mercury uptake in the surrounding environment.

A 2015 study concluded that Arctic sea ice decline accelerates methane emissions from the Arctic tundra. One of the study researchers noted, "The expectation is that with further sea ice decline, temperatures in the Arctic will continue to rise, and so will methane emissions from northern wetlands."

Atmospheric Regime

A link has been proposed between reduced Barents-Kara sea ice and cold winter extremes over northern continents. Model simulation suggest diminished Arctic sea ice may have been a contributing driver of recent wet summers over northern Europe, because of a weakened jet stream, which dives further south. Extreme summer weather in northern mid-latitudes has been linked to a vanishing cryosphere. Evidence suggest that the continued loss of Arctic sea-ice and snow cover may influence weather at lower latitudes. Correlations have been identified between high-latitude cryosphere changes, hemispheric wind patterns and mid-latitude extreme weather events for the Northern Hemisphere. A study from 2004, connected the disappearing sea ice with a reduction of available water in the American west.

Based on effects of Arctic amplification (warming) and ice loss, a study in 2015 concluded that highly amplified jet-stream patterns are occurring more frequently in the past two decades, and that such patterns can not be tied to certain seasons. Additionally it was found that these jet-stream patterns often lead to persistent weather patterns that result in extreme weather events. Hence, continued heat trapping emissions favour increased formation of extreme events caused by prolonged weather conditions.

Plant and Animal Life

Sea ice decline has been linked to boreal forest decline in North America and is assumed to culminate with an intensifying wildfire regime in this region. The annual net primary production of the Eastern Bering Sea was enhanced by 40–50% through phytoplankton blooms during warm years of early sea ice retreat.

Polar bears are turning to alternate food sources because Arctic sea ice melts earlier and freezes later each year. As a result, they have less time to hunt their historically preferred prey of seal pups, and must spend more time on land and hunt other animals. As a result, the diet is less nutritional, which leads to reduced body size and reproduction, thus indicating population decline in polar bears.

References

- Bulte, Erwin H; Joenje, Mark; Jansen, Hans GP (2000). "Is there too much or too little natural forest in the Atlantic Zone of Costa Rica?". Canadian Journal of Forest Research. 30 (3): 495–506. doi:10.1139/x99-225

- Leakey, Richard and Roger Lewin, 1996, The Sixth Extinction : Patterns of Life and the Future of Humankind, Anchor, ISBN 0-385-46809-1

- "How can you save the rain forest. 8 October 2006. Frank Field". The Times. London. 8 October 2006. Retrieved 1 April 2010

- Findell, Kirsten L.; Knutson, Thomas R.; Milly, P. C. D. (2006). "Weak Simulated Extratropical Responses to Complete Tropical Deforestation". Journal of Climate. 19 (12): 2835–2850. Bibcode:2006JCli...19.2835F. doi:10.1175/JCLI3737.1

- Taylor, Leslie (2004). The Healing Power of Rainforest Herbs: A Guide to Understanding and Using Herbal Medicinals. Square One. ISBN 9780757001444

- "Stolen Goods: The EU's complicity in illegal tropical deforestation" (PDF). Forests and the European Union Resource Network. 17 March 2015. Retrieved 31 March 2015

- Brown, Tony (1997). "Clearances and Clearings: Deforestation in Mesolithic/Neolithic Britain". Oxford Journal of Archaeology. 16 (2): 133–146. doi:10.1111/1468-0092.00030

- Ron Nielsen, The Little Green Handbook: Seven Trends Shaping the Future of Our Planet, Picador, New York (2006) ISBN 978-0-312-42581-4

- "India should follow China to find a way out of the woods on saving forest people". The Guardian. 22 July 2016. Retrieved 7 August 2016

- Pimm, S. L.; Russell, G. J.; Gittleman, J. L.; Brooks, T. M. (1995). "The Future of Biodiversity". Science. 269 (5222): 347–350. Bibcode:1995Sci...269..347P. PMID 17841251. doi:10.1126/science.269.5222.347

- Van Andel, Tjeerd H.; Zangger, Eberhard; Demitrack, Anne (2013). "Land Use and Soil Erosion in Prehistoric and Historical Greece" (PDF). Journal of Field Archaeology. 17 (4): 379–396. doi:10.1179/009346990791548628

- Mongillo, John F.; Zierdt-Warshaw, Linda (2000). Zierdt-Warshaw, Linda, ed. Encyclopedia of environmental science. University of Rochester Press. p. 104. ISBN 978-1-57356-147-1

- Rosenberg, Tina (13 March 2012). "In Africa's vanishing forests, the benefits of bamboo". New York Times. Retrieved 26 July 2012

- Pimm, S. L.; Russell, G. J.; Gittleman, J. L.; Brooks, T. M. (1995). "The future of biodiversity". Science. 269 (5222): 347–341. Bibcode:1995Sci...269..347P. PMID 17841251. doi:10.1126/science.269.5222.347

- Koop, Gary and Tole, Lise (1999). "Is there an environmental Kuznets curve for deforestation?". Journal of Development Economics. 58: 231. doi:10.1016/S0304-3878(98)00110-2

- Botkin, Daniel B. (2001). No man's garden: Thoreau and a new vision for civilization and nature. Island Press. pp. 246–247. ISBN 978-1-55963-465-6. Retrieved 4 December 2011

- Alistair B. Fraser (1994-11-27). "Bad Meteorology: The reason clouds form when air cools is because cold air cannot hold as much water vapor as warm air". Retrieved 2015-02-17

- Meyfroidt, Patrick; Lambin, Eric F. (2011). "Global Forest Transition: Prospects for an End to Deforestation". Annual Review of Environment and Resources. 36: 343. doi:10.1146/annurev-environ-090710-143732

- Prokurat, Sergiusz (2015). "Drought and water shortages in Asia as a threat and economic problem" (PDF). Journal of Modern Science. 26 (3). Retrieved 4 August 2016

- Vern Hofman; Dave Franzen (1997). "Emergency Tillage to Control Wind Erosion". North Dakota State University Extension Service. Retrieved 2009-03-21

- Culas, Richard J. (2007). "Deforestation and the environmental Kuznets curve: An institutional perspective" (PDF). Ecological Economics. 61 (2–3): 429–437. doi:10.1016/j.ecolecon.2006.03.014

- Mosley LM, Zammit B, Jolley A, and Barnett L (2014). Acidification of lake water due to drought. Journal of Hydrology. 511: 484–493

- McGuire, Thomas (2004). "Weather Hazards and the Changing Atmosphere" (PDF). Earth Science: The Physical Setting. Amsco School Pubns Inc. p. 571. ISBN 0-87720-196-X. Retrieved 2008-07-17

- Jane, Sophie (2014-09-07). "Hundreds killed by landslides and flash floods triggered by heavy monsoon rains in Kashmir". Dailymail.co.uk. Retrieved 2014-12-04

- Lam, Sai Kit; Chua, Kaw Bing (2002). "Nipah Virus Encephalitis Outbreak in Malaysia". Clinical Infectious Diseases. 34: S48–51. PMID 11938496. doi:10.1086/338818

- Mosley LM, Palmer D, Leyden E, Fitzpatrick R, and Shand P (2014). Acidification of floodplains due to river level decline during drought. Journal of Contaminant Hydrology 161: 10–23

- Mortimore, Michael (1989). Adapting to drought: farmers, famines, and desertification in west Africa. Cambridge University Press. p. 12. ISBN 978-0-521-32312-3

4

Forest Protection and Management

Forest protection is the conservation of forest regions that are under threat. Unsustainable farming, pollution of soil and urban sprawl are some of the causes of forest depletion. Forest protection and management can best be understood in confluence with the major topics listed in the following chapter. Forestry is best understood in confluence with the major topics listed in the following chapter.

Forest Protection

Forest security in Lithuania

Forest protection is the preservation or improvement of a forest threatened or affected by natural or man made causes

This forest protection also has a legal status and rather than protection from only people damaging the forests is seen to be broader and include forest pathology too. Thus due to this the different emphases around the world paradoxically suggest different things for forest protection.

In German speaking countries forest protection would focus on the biotic and abiotic factors that are non-crime related. A protected forest is not the same as a protection forest. These terms can lead to some confusion in English, although they are clearer in other languages. As a result, reading English literature can be problematic for non-experts due to localization and conflation of meanings.

The types of man induced abuse that forest protection seeks to prevent include:

- Aggressive or unsustainable farming and logging
- Pollution of soil on which forests grow
- Expanding city development caused by population explosion and the resulting urban sprawl

There is considerable debate over the effectiveness of forest protection methods. Enforcement of laws regarding purchased forest land is weak or non-existent in most parts of the world. In the increasingly dangerous South America, home of major rainforests, officials of the Brazilian National Agency for the Environment (IBAMA) have recently been shot during their routine duties.

Land Purchase

One simple type of forest protection is the purchasing of land in order to secure it, or in order to plant trees (afforestation). It can also mean forest management or the designation of areas such as natural reservoirs which are intended to be left to themselves. However, merely purchasing a piece of land does not prevent it from being used by others for poaching and illegal logging.

On Site Monitoring

A better way to protect a forest, particularly old growth forests in remote areas, is to obtain a part of it and to live on and monitor the purchased land. Even in the United States, these measures sometimes don't suffice because arson can burn a forest to the ground, leaving burnt areas free for different use.

Another issue about living on purchased forest-land is that there may not be a suitable site for a standard home without clearing land, which defies the purpose of protection. Alternatives include building a treehouse or an earthhouse. This is being done currently by indigenous people in South America to protect large reservoirs. In former times, North American Native Americans used to live in tipies or mandan earthhouses, which also require less land. An undertaking to develop modern treehouses is being taken by a company from Germany called "TrueSchool treehouses".

Methods of Protection

A number of less successful methods of forest protection have been tried, such as the trade in certified wood. Protecting a small section of land in a larger forest may also have limited value. For example, tropical rainforests can die if they decrease in size, since they are dependent on the moist microclimate which they create. There is an excellent article in National Geographic October issue concerning redwood forest in California and their effort to maintain forest and rainforest.

A compromise is to conduct agriculture and stock farming, or sustainable wood management. This ascribes different values to forest land and farmland, for which many areas are clear felled.

'Neighborhood Leakage'

Two conflicting studies on the idea that protecting forests only relocates deforestation. This is called neighborhood leakage. According to the paradox of forest protection protected areas such as rural settlements near protected zones grew at twice the rate of those elsewhere. The IUCN implements such protocols that protect over 670 eco-regions. 46% of the eco-regions had less than 10% forest protection. Which means that these areas are not being monitored as they should and the protection is not working. Considering forest protection within global priority areas was unsatisfactory. An example given was that the average protection of 8.4% in biodiversity hotspots. Results have policy relevance in terms of the target of the Convention on Biological Diversity, reconfirmed in 2008, to conserve in an effective manner that "at least 10% of each of the world's forest types".

Urban Forests

A recent discovery in Europe relating to forest protection is that urban areas have forests of their own. Many cities have tens of thousands of trees which constitute forests. In addition the air in the cities is lately becoming better, providing conditions favorable for small associated species such as mosses and lichens.

Forest Management

Forest management is a branch of forestry concerned with overall administrative, economic, legal, and social aspects, as well as scientific and technical aspects, such as silviculture, protection, and forest regulation. This includes management for aesthetics, fish, recreation, urban values, water, wilderness, wildlife, wood products, forest genetic resources, and other forest resource values. Management can be based on conservation, economics, or a mixture of the two. Techniques include timber extraction, planting and replanting of various species, cutting roads and pathways through forests, and preventing fire.

Definition

The forest is a natural system that can supply different products and services. The working of this system is influenced by the natural environment: climate, topography, soil, etc., and also by human activity. The actions of humans in forests constitute forest management. In developed societies, this management tends to be elaborate and planned in order to achieve the objectives that are considered desirable.

Some forests have been and are managed to obtain traditional forest products such as firewood, fiber for paper, and timber, with little thinking for other products and services. Nevertheless, as a result of the progression of environmental awareness, management of forests for multiple use is becoming more common.

Public Input and Awareness

There has been increased public awareness of natural resource policy, including forest management. Public concern regarding forest management may have shifted from the extraction of timber for earning money for the economy, to the preservation of additional forest resources, including wildlife and old growth forest, protecting biodiversity, watershed management, and recreation. Increased environmental awareness may contribute to an increased public mistrust of forest management professionals. But it can also lead to greater understanding about what professionals do re forests for nature conservation and ecological services. The importance of taking care of the forests for ecological as well as economical sustainable reasons has been shown in the TV show Ax Men.

Many tools like GIS and photogrammetry modelling have been developed to improve forest inventory and management planning. Since 1953, the volume of standing trees in the United States have increased by 90% due to sustainable forest management.

Wildlife Considerations

The abundance and diversity of birds, mammals, amphibians and other wildlife are affected by strategies and types of forest management.

Management Intensity

Forest management varies in intensity from a leave alone, natural situation to a highly intensive regime with silvicultural interventions. Management is generally increased in intensity to achieve either economic criteria (increased timber yields, non-timber forest products, ecosystem services) or ecological criteria (species recovery, fostering of rare species, carbon sequestration).

Forest Restoration

In the 1980s, conservation organizations warned that, once destroyed, tropical forests could never be restored. Thirty years of restoration research now challenge this: a) This site in Doi Suthep-Pui National Park, N. Thailand was deforested, over-cultivated and then burnt. The black tree stump was one of the original forest trees. Local people teamed up with scientists to repair their watershed.

b) Fire prevention, nurturing natural regeneration and planting framework tree species resulted in trees growing above the weed canopy within a year.

Forest restoration is defined as "actions to re-instate ecological processes, which accelerate recovery of forest structure, ecological functioning and biodiversity levels towards those typical of climax forest" i.e. the end-stage of natural forest succession. Climax forests are relatively stable ecosystems that have developed the maximum biomass, structural complexity and species diversity that are possible *within the limits imposed by climate and soil and without continued disturbance from humans* (more explanation here). Climax forest is therefore the target ecosystem, which defines the ultimate aim of forest restoration. Since climate is a major factor that determines climax forest composition, global climate change may result in changing restoration aims.

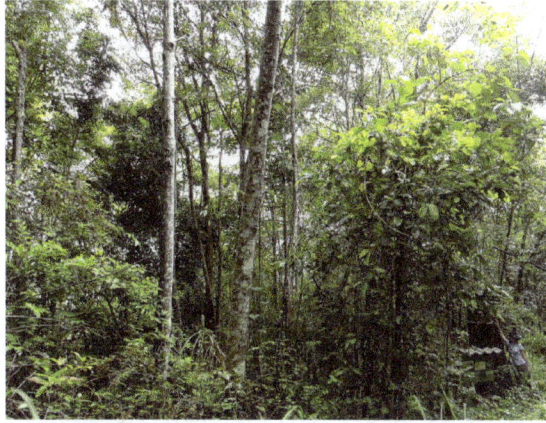

c) After 12 years, the restored forest overwhelmed the black tree stump.

Forest restoration is a specialized form of reforestation, but it differs from conventional tree plantations in that its primary goals are biodiversity recovery and environmental protection.

Scope

Forest restoration may include simply protecting remnant vegetation (fire prevention, cattle exclusion etc.) or more active interventions to accelerate natural regeneration, as well as tree planting and/or sowing seeds (direct seeding) of species characteristic of the target ecosystem. Tree species planted (or encouraged to establish) are those that are typical of, or provide a critical ecological function in, the target ecosystem. However, wherever people live in or near restoration sites, restoration projects often include economic species amongst the planted trees, to yield subsistence or cash-generating products.

Forest restoration is an inclusive process, which depends on collaboration among a wide range of stakeholders including local communities, government officials, non-government organizations, scientists and funding agencies. Its ecological success is measured in terms of increased biological diversity, biomass, primary productivity, soil organic matter and water-holding capacity, as well as the return of rare and keystone species, characteristic of the target ecosystem. Economic indices of success include the value of forest products and ecological services generated (e.g. watershed protection, carbon storage etc.), which ultimately contribute towards poverty reduction. Payments for such ecological services (PES) and forest products can provide strong incentives for local people to implement restoration projects.

Opportunities for Forest Restoration

Forest restoration is appropriate wherever biodiversity recovery is one of the main goals of reforestation, such as for wildlife conservation, environmental protection, eco-tourism or to supply a wide variety of forest products to local communities. Forests can be restored in a wide range of circumstances, but degraded sites within protected areas are a high priority, especially where some climax forest remains as a seed source within the landscape. Even in protected areas, there are often large deforested sites: logged over areas or sites formerly cleared for agriculture. If protected areas are to act as Earth's last wildlife refuges, restoration of such areas will be needed.

Demonstration forest restoration plot, SUNY-ESF, Syracuse, NY

Many restoration projects are now being implemented under the umbrella of "forest landscape restoration" (FLR), defined as a "planned process to regain ecological integrity and enhance human well-being in deforested or degraded landscapes". FLR recognizes that forest restoration has social and economic functions. It aims to achieve the best possible compromise between meeting both conservation goals and the needs of rural communities. As human pressure on landscapes increases, forest restoration will most commonly be practiced within a mosaic of other forms of forest management, to meet the economic needs of local people.

Natural Regeneration

Tree planting is not always essential to restore forest ecosystems. A lot can be achieved by studying how forests regenerate naturally, identifying the factors that limit regeneration and devising methods to overcome them. These can include weeding and adding fertilizer around natural tree seedlings, preventing fire, removing cattle and so on. This is "accelerated" or "assisted" natural regeneration. It is simple and cost-effective, but it can only operate on trees that are already present, mostly light-loving pioneer species. Such tree species are not usually those that comprise climax forests, but they can foster recolonization of the site by shade-tolerant climax forest tree species, via natural seed dispersal from remnant forest. Because this is a slow process, biodiversity recovery can usually be accelerated by planting some climax forest tree species, especially large-seeded, poorly dispersed species. It is not feasible to plant all the tree species that may have formerly grown in the original primary forest and it is usually unnecessary to do so, if the framework species method can be used.

Post-fire Regeneration

In large parts of the world, forest fires cover a heavy toll on forests. That can be because of provoked deforestation in order to substitute forests by crop areas, or in dry areas, because of wild fires occurring naturally or intentionally. A whole section of forest landscape restoration in linked to this particular problem, as in many cases, the net loss of ecosystem value is very high and can open the drop to an accelerated further degradation of the soil conditions through erosion and desertification. This indeed has dire consequences on both the quality of the habitats and their related fauna. Nevertheless, in some specific cases, wild fires do actually allow to increase the bio-diversity index of the burnt area, in which case the Forest Restoration Strategies tend to look for a different land-use.

Forest Restoration Projects

Ashland Forest Resiliency Stewardship Project

The Ashland Forest Resiliency Stewardship Project (AFR) is a decade long, science-based project launched in 2010 with the intent of reducing severe wildfire risk, but also protecting water quality, old-growth forest, wildlife, people, property, and the overall quality of life within the Ashland watershed. The primary stakeholders in this cooperative restoration effort are the U.S. Forest Service, the City of Ashland, Lomaktsi Restoration Project, and the Nature Conservancy. The project was launched with initial funding from the Economic Recovery stimulus, and has more recently received funding from the Forest Service Hazardous Fuels program and the Joint Chiefs Landscape Restoration Partnerships program to back the project through 2016 .

Located in the dry forests of southern Oregon, the threat of wildfire is a reality for land managers and property owners alike. The boundaries of the city of Ashland intersect with the surrounding forest in what is referred to as the wildland-urban interface (WUI). Historically, the forests of this region experienced a relatively frequent fire return interval, which prevented buildup of heavy fuel loads. A century of fire exclusion and suppression on federal lands in the Pacific Northwest has led to increased forest density and fuel loads, and thus a more persistent threat of devastating wildfire.

The AFR project has implemented restoration techniques and prescriptions that aim to replicate the process of ecological succession in dry, mixed-conifer forests of the Pacific Northwest. The approach involves a combination of fuels reduction, thinning small-diameter trees, and carrying out prescribed burns. Priority is given to maintaining ecological function and complexity by retaining the largest and oldest trees, preserving wildlife habitat and riparian areas, and protecting erodible soils and maintaining slope stability.

Since its inception in 2010, the AFR project has completed restoration work on 4,000 of the 7,600 acres slated for the project. The project has provided educational experience to over 2,000 students and has benefitted the local community by creating jobs and providing workforce training. Currently, helicopter logging operations are thinning 1,100 acres of the watershed while controlled burning operations take place as air quality conditions allow.

Forest Landscape Restoration

Forest landscape restoration (FLR) is defined as "a planned process to regain ecological integrity and enhance human well-being in deforested or degraded landscapes". It comprises tools and procedures to integrate site-level forest restoration actions with desirable landscape-level objectives, which are decided upon via various participatory mechanisms among stakeholders. The concept has grown out of collaboration among some of the world's major international conservation organizations including the International Union for Conservation of Nature (IUCN), the World Wide Fund for Nature (WWF), the World Resources Institute and the International Tropical Timber Organization (ITTO).

Aims

The concept of FLR was conceived to bring about compromises between meeting the needs of both humans and wildlife, by restoring a range of forest functions at the landscape level. It includes actions to strengthen the resilience and ecological integrity of landscapes and thereby keep future management options open. The participation of local communities is central to the concept, because they play a critical role in shaping the landscape and gain significant benefits from restored forest resources. Therefore, FLR activities are inclusive and participatory.

Desirable Outcomes

The desirable outcomes of an FLR program usually comprise a combination of the following, depending on local needs and aspirations:

- identification of the root causes of forest degradation and prevention of further deforestation,

- positive engagement of people in the planning of forest restoration, resolution of land-use conflicts and agreement on benefit-sharing systems,

- compromises over land-use trade-offs that are acceptable to the majority of stakeholders,

- a repository of biological diversity of both local and global value,

- delivery of a range of utilitarian benefits to local communities including:

  o a reliable supply of clean water,

  o environmental protection particularly watershed services (e.g. reduced soil erosion, lower landslide risk, flood/drought mitigation etc.),

  o a sustainable supply of a diverse range of forest products including foods, medicines, firewood etc.,

  o monetary income from various sources e.g. ecotourism, carbon trading via the REDD+ mechanism and from payments for other environmental services (PES)

Activities

FLR combines several existing principles and techniques of development, conservation and natural resource management, such as landscape character assessment, participatory rural appraisal, adaptive management etc. within a clear and consistent evaluation and learning framework. An FLR program may comprise various forestry practices on different sites within the landscape, depending on local environmental and socioeconomic factors. These may include protection and management of secondary and degraded primary forests, standard forest restoration techniques such as "assisted" or "accelerated" natural regeneration (ANR) and the planting of framework tree species to restore degraded areas, as well as conventional tree plantations and agroforestry systems to meet more immediate monetary needs.

The IUCN hosts the The Global Partnership on Forest Landscape Restoration, which co-ordinates development of the concept around the world.

In 2014, the Food and Agricultural Organization of the United Nations established the Forest and Landscape Restoration Mechanism. The Mechanism supports countries to implement FLR as a contribution to achieving the Bonn Challenge - the restoration of 150 million hectare of deforested and degraded lands by 2020 - and the Convention on Biological Diversity Aichi Biodiversity Targets - related to ecosystem conservation and restoration.

In partnership with the Global Mechanism of the United Nations Convention to Combat Desertification, FAO released two discussion papers on sustainable financing for FLR in 2015. *Sustainable Financing for Forest and Landscape Restoration: The Role of Public Policy Makers* provides recommendations and examples of FLR financing for countrues. *Sustainable Financing for Forest and Landscape Restoration - Opportunities, challenges and the way forward* provides an overview of funding sources and financial instruments available for FLR activities.

Forest Inventory

Forest inventory is the systematic collection of data and forest information for assessment or analysis. An estimate of the value and possible uses of timber is an important part of the broader information required to sustain ecosystems. When taking forest inventory the following are important things to measure and note: species, diameter at breast height (DBH), height, site quality, age, and defects. From the data collected one can calculate the number of trees per acre, the basal area, the volume of trees in an area, and the value of the timber. Inventories can be done for other reasons than just calculating the value. A forest can be cruised to visually assess timber and determine potential fire hazards and the risk of fire. The results of this type of inventory can be used in preventative actions and also awareness. Wildlife surveys can be undertaken in conjunction with timber inventory to determine the number and type of wildlife within a forest. The aim of the statistical forest inventory is to provide comprehensive information about the state and dynamics of forests for strategic and management planning. Merely looking at the forest for assessment is called taxation.

History

Surveying and taking inventory of trees originated in Europe in the late 18th century out of a fear that wood (the main source of fuel) would run out. The first information was organized into maps used to plan out usage. In the early 19th century forest harvesters estimated the volume and dispersal of trees within smaller forests with their eyes. More diverse and larger forests were divided into smaller sections of similar type trees that were individually estimated by visual inspection. These estimates were related together to figure out the entire forest's available resources. As the 19th century progressed so did the measurement techniques. New relationships between diameter, height, and volume were discovered and exploited. These newfound relationships allowed for a more accurate assessment of wood types and yields of much larger forests. By 1891, these surveys were conducted through sample-based methods involving statistical averages and more sophisticated measuring devices were implemented. In the 20th century, the statistical method of sampling had become well established and commonly used. Further developments, such as unequal probability sampling, arose. As the 20th century progressed, an understanding of co-efficients of error became clearer and the new technology of computers combined with the availability of aerial as well as satellite photography, further refined the process. Laser Scanning both terrestrially and aerially are now used alongside more manual methods. As a result, sampling accuracy and assessment values became more accurate and allowed for modern practices to arise.

A forest inventory does not only record the trees height, DBH and number for tree yield calculations. But it also records the conditions of the forest. So this would include, geology, site conditions, tree health and other forest factors.

Timber Cruise

A timber cruise is a sample measurement of a stand used to estimate the amount of standing timber that the forest contains. These measurements are collected at sample locations called plots, quadrants, or strips. Each of these individual sample areas is one observation in a series of observations called a sample. These sample areas are generally laid out in some random fashion usually in the form of a line plot survey. Depending on the size of the plot and the number of plots measured, the data gathered from these plots can then be manipulated to achieve varying levels of certainty for an estimate that can be applied to the entire timber stand. This estimate of stand conditions, species composition, volume and other measured attributes of a forest system can then be used for various purposes. For example, in British Columbia the sale of Crown timber is a business proposition and both the buyer and the Ministry of Forests and Range (seller) must know the quantity and the quality of timber being sold. Our satellite provides the essential data and information for determining stumpage rates, for establishing conditions of sale and for planning of the logging operations by the licensee. Generally a timber cruise includes measurements or estimates of timber volume by forest product sort (and sometimes grade), log defect, and log lengths, whether the estimates are made in the field or using computer software.

Stranded Examinations

Before, the most common type of inventory is one that uses a random sampling technique which groups similar forests into one category based on age, stand structure, species, and location. The next step would be to begin from a random place and measure circular plots that are equidistant from one another. Inventory undertaken in conjunction with wildlife surveys has a regularized distance between plots. When taking inventory, a decision has to be made about which types of plots to measure. There are several different types of plots. The two most common types are *fixed radius* and *variable radius*, also known as *prism radius*. In a fixed radius plot, the forester finds the center of a plot and every tree within a certain fixed distance away from that point is measured. Variable radius plots are used more for inventory of volume. During this method, an angle is created and projected from the plot center and all trees that are larger than the projected right angle are measured with a ruler called a Tunagmetor .

Example of hardware equipment for forest inventories: GPS and laser rangefinder for mapping connected to a field rugged computer.

Types of Sample Plots

Fixed Area Plot

Fixed area plot sample measurements are taken so that they are a fraction of the entire timber stand. This means that the numbers are all proportional to the actual stand values and that by multiplying by the correct corresponding value you can obtain the actual tract values. These plots are taken randomly so that each sample point has an equal probability of being included in the random sample. Commonly a 100m² plot is taken.

Variable Size Plots

A variable size plot is more dependent of the size of the trees. The tract is measured on a series of points and the trees are tallied for being in or out depending on their size and location relevant to the plot center. Usually an angle gauge, wedge prism or Relascope are used to gather data for this type of plot. This allows for a very quick estimate of the volume and species of a given tract.

Transects

Transects are arbitrarily determined lines (to prevent sampling bias) through a stand employed as a linear form of sample plot. They are sometimes referred to as "strip lines."

Plot Selection

Plots are samples of the forest being inventoried and so are selected according to what is looked for.

Simple Random Sampling

A computer or calculator random number generator is used to assign plots to be sampled. Here random means an equal chance of any plot being selected out of all of the plots available. It does not mean haphazard. Often it is modified to avoid sampling roads, ensure coverage of unsampled areas and for logistics of actually getting to the plots.

Systematic Sampling

Commonly this is done by a random point and then laying a grid over a map of the area to be sampled. This grid will have preassigned plot areas to be sampled. It means more efficient logistics and removes some of the human bias that may be there with simple random sampling.

Systematic Stratified Sampling

There is some HAPPY grouping, for example by age classes or soil characteristics or slope elevation. And then plots are chosen from each grouping by another sampling technique. It requires some knowledge of the land first and also trust that the groupings have been done properly. In forestry it may be done to separate plantation areas from mixed forest for example and reduce the amount of sampling time needed.

Systematic Clustered Sampling

When it is not possible to make strata for stratified sampling, there may be some knowledge about the forest where it can be said that small groupings are possible. These small groupings of plots if they are near to each other form a cluster. These clusters are then randomly sampled with the belief that they are representing the actual mix of the forest. As they are close to each other there is less waking needed and so it is more efficient.

Timber Metrics

The amount of standing timber that a forest contains is determined from:

- Age class. This is a misnomer as in German it is Wuchsklass (Growth class) and properly should be the size of the tree (Size class), which may be limited by shade over it and not necessarily the biological and thus physiological age of the tree. This distinction is important if tree growth over time is expected by an owner or forester to produce timber, as a small old tree will grow differently from a small young tree. Commonly these age classes are split into: Seedling, Sapling, Pole, Mature Tree (subdivided into Weak wood, Middle wood and Strong wood stages), Old / Scenescant Tree. Sometimes it is called size class or a cohort. There are differences between countries and forests.

- Basal area – defines the area of a given section of land that is occupied by the cross-section of tree trunks and stems at their base

- Diameter at breast height (DBH) – measurement of a tree's girth standardized with different countries having different standards they are often at 1.3 meters (about 4.5 feet) above the ground

- Form factor – the shape of the tree, based on recorded trees and commonly then given for calculating tree volumes for a given species. It is usually related to DBH or age class. It is distinct from taper. So it can be conical or paraboloid for example.

- Girard form class – an expression of tree taper calculated as the ratio of diameter inside the bark at 16 feet above ground to that outside bark at DBH, primary expression of tree form used in the United States

- Quadratic mean diameter – diameter of the tree that coordinates to the stand's basal area

- Site index – a species specific measure of site productivity and management options, reported as the height of dominant and co-dominant trees (site trees)in a stand at a base age such as 25, 50 and 100 years

- Tree taper – the degree to which a tree's stem or bole decreases in diameter as a function of height above ground. So it can be sharp or gradual.

Volume Estimation

- Stocking – a quantitative measure of the area occupied by trees relative to an optimum or desired level of density, which will vary according to management purpose

- Stand Density Index – a measure of the stocking of a stand of trees based on the number of trees per unit area and DBH of the tree of average basal area, based on historical records and is particular to site type and species

- Volume table – a chart based on volume equations that uses correlations between certain aspects of a tree to estimate the standing volume

- Stand density management diagram – model that uses current stand density to project future stand composition

Volume can be calculated from the metrics recorded in a plot sample. For example, if a tree was measured to be 20m tall and with a DBH of 19 cm using previous measured tree data a volume could be approximated according to species. Such a table has been constructed by Josef Pollanschütz in Austria.

Volume of tree = BA X h x f pollanschutz

So f pollanschütz would be derived from the table and is properly called the Form Factor.

To scale this up to a hectare level the result would have to be multiplied by the number of trees of that size. This is called the blow up factor.

Tools Used in Inventory

- Biltmore stick – utilizes ocular trigonometry to quickly measure diameter and height

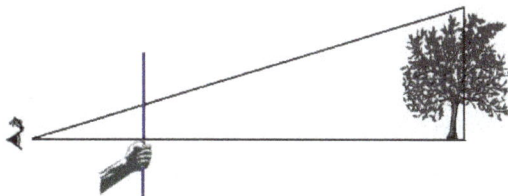

Figure demonstrating the ocular trigonometric principles behind the Biltmore stick.

- Diameter tape – cloth or metal tape that is wrapped around the bole, scaled to diameter

- Caliper – two prongs connected to a measuring tape are placed around the most average part of the bole to determine diameter

- Relascope – multiple-use tool that is able to find tree height, basal area, and tree diameter anywhere along the bole

- Clinometer – common tool used to measure changes in elevation and tree height

- Cruising rod – similar to a caliper, calculates the number of pieces of lumber yielded by a given piece of timber by measuring its diameter

- LASER Scanner used with computer software to calculate the metrics from the collected data by use of Lidar.

- Wedge prism – a small glass wedge that refracts light to create visible offsets in order to be able to choose which trees at a sampling point should be included in the sample.

- Data collector – an electronic device used to quickly enter sample data, geo-locate the data, and, in more modern times, to also access reference, web and historic materials while timber cruising.

- Increment borer – a device used to retrieve a cylindrical sample of wood material orthogonally from the stem while doing as little damage as possible to the remaining tissues.

In 2014, the Food and Agriculture Organization of the United Nations and partners, with the support of the Government of Finland, launched Open Foris – a set of open-source software tools that assist countries in gathering, producing and disseminating reliable information on the state of forest resources. The tools support the entire inventory lifecycle, from needs assessment, design, planning, field data collection and management, estimation analysis, and dissemination. Remote sensing image processing tools are included, as well as tools for international reporting for REDD+ MRV and FAO's Global Forest Resource Assessments.

References

- Schmitt, C.; Burgess, N. (2009). "Global analysis of the protection status of the world's forests". Biological Conservation. 142 (10). doi:10.1016/j.biocon.2009.04.012

- Shindler, Bruce; Lori A. Cramer (January 1999). "Shifting Public Values for Forest Management: Making Sense of Wicked Problems". Western Journal of Applied Forestry. Society of American Foresters. 14 (1): 28–34. ISSN 0885-6095. Retrieved 2008-08-25

- "City of Ashland, Oregon - Ashland Forest Resiliency Project - AFR Work Update". www.ashland.or.us. Retrieved 2016-02-05

- Regos, Adrian; D'Amen, M; Titeux, Nicolas; Herrando, Sergi; Guisan, A; Brotons, Lluis (2016). "Predicting the future effectiveness of protected areas for bird conservation in Mediterranean ecosystems under climate change and novel fire regime scenarios". Diversity and Distributions. 22 (1): 83–96. doi:10.1111/ddi.12375

- Toman, E., Stidham, M., Shindler, B., McCaffrey, S. 2011. Reducing fuels in the wildland-urban interface: community perceptions of agency fuel treatments. Intl. Journal of Wildland Fire 20 (3):340-349

- Nations, Food and Agriculture Organization of the United. "Assisted natural regeneration of forests". www.fao.org. Retrieved 2016-02-24

- Martín-Alcón, Santiago; Coll, Lluis (2016). "Unraveling the relative importance of factors driving post-fire regeneration trajectories in non-serotinous Pinus nigra forests" (PDF). Forest Ecology and Management. 361: 13–22. doi:10.1016/j.foreco.2015.11.006

- Sustainable Financing for Forest and Landscape Restoration - Opportunities, challenges and the way forward

(PDF). Food and Agriculture Organization of the United Nations and the Global Mechanism of the UNCCD. June 2015. ISBN 978-92-5-108992-7

- Franklin, J.F., and Johnson, K.N. 2012. A restoration framework for federal forests in the Pacific Northwest. Journal of Forestry 110 (8): 429-439

- "Glossary of Forestry Terms in British Columbia" (pdf). Ministry of Forests and Range (Canada). March 2008. Retrieved 2009-04-06

- L. J. Moores; B. Pittman; G. Kitchen (1996), "Forest ecological classification and mapping: their application for ecosystem management in Newfoundland", Environmental Monitoring and Assessment, 39 (1–3): 571–577, doi:10.1007/bf00396169

Permissions

All chapters in this book are published with permission under the Creative Commons Attribution Share Alike License or equivalent. Every chapter published in this book has been scrutinized by our experts. Their significance has been extensively debated. The topics covered herein carry significant information for a comprehensive understanding. They may even be implemented as practical applications or may be referred to as a beginning point for further studies.

We would like to thank the editorial team for lending their expertise to make the book truly unique. They have played a crucial role in the development of this book. Without their invaluable contributions this book wouldn't have been possible. They have made vital efforts to compile up to date information on the varied aspects of this subject to make this book a valuable addition to the collection of many professionals and students.

This book was conceptualized with the vision of imparting up-to-date and integrated information in this field. To ensure the same, a matchless editorial board was set up. Every individual on the board went through rigorous rounds of assessment to prove their worth. After which they invested a large part of their time researching and compiling the most relevant data for our readers.

The editorial board has been involved in producing this book since its inception. They have spent rigorous hours researching and exploring the diverse topics which have resulted in the successful publishing of this book. They have passed on their knowledge of decades through this book. To expedite this challenging task, the publisher supported the team at every step. A small team of assistant editors was also appointed to further simplify the editing procedure and attain best results for the readers.

Apart from the editorial board, the designing team has also invested a significant amount of their time in understanding the subject and creating the most relevant covers. They scrutinized every image to scout for the most suitable representation of the subject and create an appropriate cover for the book.

The publishing team has been an ardent support to the editorial, designing and production team. Their endless efforts to recruit the best for this project, has resulted in the accomplishment of this book. They are a veteran in the field of academics and their pool of knowledge is as vast as their experience in printing. Their expertise and guidance has proved useful at every step. Their uncompromising quality standards have made this book an exceptional effort. Their encouragement from time to time has been an inspiration for everyone.

The publisher and the editorial board hope that this book will prove to be a valuable piece of knowledge for students, practitioners and scholars across the globe.

Index